OUR CHANGING PLANET

"Climate change represents one of our greatest challenges of our time, but it is a challenge uniquely suited to America's strengths…the Administration has continued, through the U.S. Global Change Research Program, to support science and monitoring to expand our understanding of climate change and its impacts."

– The President's Climate Action Plan, June 2013

This material was developed with Federal support through the U.S. Global Change Research Program under National Science Foundation Award No. NSFDACS13C1421 (Cooperative Agreement No. AGS-0936594)

U.S. Global Change Research Program, Suite 250, 1717 Pennsylvania Ave, NW, Washington, DC 20006
Tel: +1 202 223 6262 | Fax: +1 202 223 3065

Subcommittee on Global Change Research

CHAIR

Thomas Karl
Department of Commerce

VICE CHAIRS

Jeffrey Arnold
Department of Defense
Vice Chair, Adaptation Science

Mike Freilich
National Aeronautics and Space Administration
Vice Chair, Integrated Observations

Gerald Geernaert
Department of Energy
Vice Chair, Integrated Modeling

Roger Wakimoto
National Science Foundation
Vice Chair, Process Research

PRINCIPALS

John Balbus
Department of Health and Human Services

Katharine Batten
U.S. Agency for International Development

Joel Clement
Department of the Interior

Robert Detrick
Department of Commerce

Scott Harper, Acting
Department of Defense

Leonard Hirsch
Smithsonian Institution

William Hohenstein
Department of Agriculture

Jack Kaye
National Aeronautics and Space Administration

Michael Kuperberg
Department of Energy

C. Andrew Miller
Environmental Protection Agency

Joann Roskoski
National Science Foundation

Arthur Rypinski, Acting
Department of Transportation

Trigg Talley
Department of State

EXECUTIVE OFFICE AND OTHER LIAISONS

Thomas Armstrong
Executive Director
U.S. Global Change Research Program
White House Office of Science and Technology Policy

Tamara Dickinson
Principal Assistant Director for Environment and Energy
White House Office of Science and Technology Policy

Richard Duke
Associate Director for Energy and Climate Change
White House Council on Environmental Quality

Fabien Laurier
Director, Third National Climate Assessment
White House Office of Science and Technology Policy

Kimberly Miller
Program Examiner
White House Office of Management and Budget

Christopher Weaver
Deputy Director
U.S. Global Change Research Program

FISCAL YEAR 2014

June 2014

Members of Congress:

On behalf of the National Science and Technology Council, I am pleased to transmit *Our Changing Planet: The U.S. Global Change Research Program for Fiscal Year 2014*. The U.S. Global Change Research Program (USGCRP) coordinates scientific research across 13 Federal departments and agencies whose missions include understanding changes in the global environment and their implications for society. In accordance with the Global Change Research Act of 1990 (GCRA), the enclosed report summarizes USGCRP's recent achievements, current status, future priorities, and FY 2014 budget information.

In June 2013, the President released his Climate Action Plan, laying out the case for action on climate change and the steps his Administration will take to address it, and drawing attention to USGCRP's important role in informing this process. In this new edition of *Our Changing Planet*, USGCRP will illustrate how accomplishments and investments that support its 2012–2021 Strategic Plan—as well as the President's Climate Action Plan—build upon current strengths in integrated observations, research, modeling, and information services to serve societal needs. A brief exploration of the essential research infrastructure of USGCRP will lead into highlighted examples of key societally relevant program focus areas. Sample cases—such as extreme weather and climate events, rising sea levels and melting ice, and climate-change impacts on human health—will exemplify the end-to-end nature of USGCRP that progresses from research to sustained assessment, education, communication, and decision support. This approach fully addresses the GCRA mandate to *"understand, assess, predict, and respond to human-induced and natural processes of global change."*

This *Our Changing Planet* report summarizes USGCRP's significant progress toward achieving its strategic goals, supporting the President's Climate Action Plan, and building a knowledge base that effectively informs human responses to global change. I appreciate the close cooperation of the participating agencies and look forward to working with members of the Congress to implement the continuation of this essential National program.

Sincerely,

Dr. John P. Holdren
Director, Office of Science and Technology Policy
Assistant to the President for Science and Technology

iii

Since 1989, the U.S. Global Change Research Program (USGCRP) has submitted annual reports to Congress called *Our Changing Planet*. The reports describe the status of USGCRP research activities, provide progress updates, and document recent accomplishments. This FY 2014 edition of *Our Changing Planet* provides a summary of programmatic achievements, recent progress, future priorities, and budgetary information. It thereby meets the requirements set forth in the U.S. Global Change Research Act of 1990 (Section 102, P. L. 101–606) to provide an annual report on Federal global change research priorities and programs. It does not express any regulatory policies of the United States or any of its agencies, or make any findings that could serve as predicates for regulatory action.

Photo Credits:

front and back cover: NASA
page vi: Public domain
page 2: University Corporation for Atmospheric Research (Carlye Calvin)
page 4: Public domain
page 6: University Corporation for Atmospheric Research (Carlye Calvin)
page 20: University Corporation for Atmospheric Research (James Hannigan)
page 28: University Corporation for Atmospheric Research (Carlye Calvin)
page 30: University Corporation for Atmospheric Research (Carlye Calvin)

Table of Contents

INTRODUCTION

In 1990, Congress passed the U.S. Global Change Research Act (GCRA of 1990, Section 102, P.L. 101–606), creating the U.S. Global Change Research Program (USGCRP) with a mandate to *"assist the Nation and the world to understand, assess, predict, and respond to human-induced and natural processes of global change."*[1]

More than two decades later, USGCRP continues to play a pivotal role in coordinating and integrating scientific research, information, tools, and technologies across the Federal Government. In June 2013, President Obama launched a comprehensive Climate Action Plan that called for sound science to manage the impacts of climate change.[2] Many elements of the Plan—including Developing Actionable Science, Assessing Climate Change Impacts in the United States, Launching a Climate Data Initiative, and Providing a Toolkit for Climate Resilience—build on the tenets of the GCRA and will be supported by USGCRP's ongoing and planned investments in science, assessment, and decision support.

As a whole, the GCRA, the President's Climate Action Plan, and the Federal Government's continued investment in USGCRP mark an important and enduring recognition that global change is happening and that action is needed to understand and address the increasingly severe climate-change impacts affecting our Nation. 2012 was the hottest year ever in the contiguous United States. Globally, nine of the 10 hottest years on record have occurred in the 21st century. Many communities across the Nation are experiencing longer and hotter summers, shorter and warmer winters, and record-breaking periods of extreme heat. Because of rising sea levels, residents of coastal cities are experiencing more frequent floods from storm surges. Hotter and drier weather and earlier snow melt in the western United States mean that wildfires start earlier in the year,

occur later into the fall, and burn more acreage—threatening lives, homes, and ecosystems. In Alaska, the summer sea ice that once protected coasts has receded, and fall storms are causing erosion and damage severe enough that some communities are facing relocation.

Scientists around the world have demonstrated that these observations are consistent with global climatic trends. Long-term, independent records from weather stations, satellites, ocean buoys, tide gauges, and many other data sources all confirm that the United States and the world are warming, that precipitation patterns are changing, that global average sea level is rising, and that some types of extreme weather events are becoming more frequent and more severe. These changes have resulted in a wide range of impacts across every region of our Nation and many sectors of our economy. Climate change is no longer a distant threat—it carries immediate and long-term consequences, as well as economic costs.

In light of these changes, Americans across the country are facing choices about how to plan for and address climate-related impacts that affect their businesses, communities, and families. Corn producers in Iowa, oyster farmers in Washington, and maple syrup producers in Vermont, for example, have all observed changes to their local climates that are unprecedented in their experience. So, too, have coastal planners from Florida to Maine, water managers in the arid Southwest, and public health officials in every region of the Nation. These and many other stakeholders need reliable tools and scientific information about current and future climate changes, impacts, and effective response options in order to do their jobs and make informed decisions. Over the past year, USGCRP continued its fundamental work towards fulfilling this urgent and critical need.

---

[1] **Website:** www.globalchange.gov/about/legal-mandate

[2] **Website:** whitehouse.gov/sites/default/files/image/president27sclimate actionplan.pdf

2 USGCRP'S VISION, MISSION, AND MANDATE

USGCRP works to fulfill the mandate of the GCRA by coordinating the Federal Government's $2.5 billion annual investment in global change research—the largest such investment in the world (*see Box 1. USGCRP's Vision and Mission*). Since its inception, USGCRP has supported a Federal global change research enterprise that provides taxpayers substantial returns on this investment, including major **advances in our knowledge of Earth's past and present climate, improved climate-change projections for the future, and a better understanding of society's vulnerabilities to the impacts of global change.**

The science portfolio coordinated by USGCRP spans large scales of space and time, and includes changes being wrought by human behavior as well as by natural drivers. It incorporates nearly all forms of scientific work, including laboratory experiments, field research, computer modeling, synthesis and assessment, and observations of Earth from land, air, sea, and space. This vast body of work is carried out by 13 Federal agencies, each with its own mission and areas of expertise (*see Box 2. USGCRP Agencies*).

USGCRP provides both a skeletal framework for and connective tissue between these member agencies so that together, they produce more effective, efficient, and holistic results. This is accomplished in a variety of ways—by developing joint research priorities, enabling knowledge and capacity transfer between agencies, minimizing redundancy across projects, and leveraging distributed Federal resources—an especially important task during these austere budget times.

The scientific knowledge base that USGCRP is continually expanding serves an important foundation for informed decision making. USGCRP-enabled science activities that support an improved understanding of and preparedness for current and future global changes include:

- *Observing Changes in the Earth System:* Investing in observations of the Earth system, including satellite observations, that allow scientists to monitor global change and understand climate processes;

- *Understanding the Complex Earth System and Advancing Use-Inspired Science for Adaptation and Mitigation:* Conducting fundamental and use-inspired research to better understand and respond to changes in the integrated Earth system;

- *Modeling Global Changes:* Developing, testing, and applying sophisticated models—the principal tools scientists use to project future climate;

- *Conducting and Sustaining Domestic and International Global Change Assessments:* Assessing climate changes and related impacts in the United States and the world by synthesizing available scientific information from peer-reviewed literature and other credible sources;

- *Understanding Global Change Preparedness, Adaptation, and Resilience:* Researching

---

Box 1. USGCRP's Vision & Mission

Vision – A Nation, globally engaged and guided by science, meeting the challenges of climate and global change.

Mission – To build a knowledge base that informs human responses to climate and global change through coordinated and integrated Federal programs of research, education, communication, and decision support.

preparedness, adaptation, and resilience while supporting the needs of decision makers, in part through increased engagement between scientists and information users;

- *Sharing and Managing Critical Global Change Data and Information:* Sharing and managing data, information, and user-friendly tools for access, analysis, and knowledge transfer;

- *Engaging, Educating, and Communicating with Stakeholders:* Communicating scientific findings to diverse audiences, through engaging with and educating the public, members of Congress, and the global research community.

In April 2012, USGCRP released its National Global Change Research Plan 2012–2021: A Strategic Plan for the U.S. Global Change Research Program,[3] laying out clear goals to achieve an ambitious new vision, including the expansion of stakeholder participation in the scientific process and the dissemination of results and information to broad audiences (*see Box 3. USGCRP's Four Strategic Plan Goals*). The Strategic Plan emphasizes fundamental and use-inspired research that can benefit society, creating the knowledge base to answer critical questions about the changing Earth system and how we can respond. Such research requires multiple forms of integration: across natural and human systems, space and time, observations and modeling, scientific capabilities and stakeholder needs, and domestic and international partnerships.

A substantial amount of work is underway at USGCRP to achieve the vision outlined in its Strategic Plan, as well as in the President's Climate Action Plan—with many notable successes already achieved. The following sections of this document review the major components of the USGCRP research enterprise; examples of recent accomplishments and investments, both in advancing the science of global change and in synthesizing, communicating, and applying that science for decision making; and an outlook on program priorities for Fiscal Year (FY) 2014. Budget information supporting this narrative is provided at the end of the report. The activities highlighted in this document demonstrate foundational research accomplishments and needs for delivering on the GCRA mandate and addressing critical components of the President's Climate Action Plan.

Box 2. USGCRP Agencies

- Department of Agriculture (USDA)
- Department of Commerce (DOC)
- Department of Defense (DOD)
- Department of Energy (DOE)
- Department of Health and Human Services (HHS)
- Department of the Interior (DOI)
- Department of State (DOS)
- Department of Transportation (DOT)
- Environmental Protection Agency (EPA)
- National Aeronautics and Space Administration (NASA)
- National Science Foundation (NSF)
- Smithsonian Institution (SI)
- U.S. Agency for International Development (USAID)

Box 3. USGCRP's Four Strategic Plan Goals

Goal 1 – Advance Science: Advance scientific knowledge of the integrated natural and human components of the Earth system.

Goal 2 – Inform Decisions: Provide the scientific basis to inform and enable timely decisions on adaptation and mitigation.

Goal 3 – Conduct Sustained Assessments: Build sustained assessment capacity that improves the Nation's ability to understand, anticipate, and respond to global change impacts and vulnerabilities.

Goal 4 – Communicate and Educate: Advance communication and education to broaden public understanding of global change and develop the scientific workforce of the future.

---

[3] **Website:** downloads.globalchange.gov/strategic-plan/2012/usgcrp-strategic-plan-2012.pdf

3 FEDERAL INVESTMENTS IN GLOBAL CHANGE RESEARCH

3.1 Understanding Current and Future Changes in the Global Climate

Observing Changes in the Earth System

USGCRP research depends upon a comprehensive, integrated, and continuous set of physical, chemical, biological, and societal observations of global change and its impacts. Raw observations are collected remotely and *in situ* across Earth's atmosphere, oceans, land, and ice using a variety of high-tech instruments, and are transformed into value-added products through data analysis and integration. USCGRP's portfolio of Earth observations includes satellite, airborne, ground-based, and ocean-based missions, platforms, and networks—all of which provide measurements of Earth system variables that are important for understanding and responding to global change (*see Box 4. Continuing Critical Earth Observations with Landsat Satellites*).

In the coming years, USGCRP's Earth observations portfolio will continue to focus on collecting data needed to document, understand, and respond to changing climate patterns and long-term Earth system trends, as well as on data needed for regional-scale planning and decision making (*see Box 5. Supporting Global Change Research Through Satellite Missions: A Look Ahead*). One area of emphasis going forward will be to clarify changing patterns in the magnitude, frequency, and distribution of extreme events such as wildfires, tornadoes, severe thunderstorms, lightning, and hurricanes, among others.

Understanding the Complex Earth System

Using integrated observations of the Earth system, as well as sophisticated models that capture the important intersections of the physical climate system and life on Earth, USGCRP is continuing to advance Earth system science and grow the knowledge base needed to manage the risks associated with global change. For example,

several USGCRP agencies are investigating the history of Earth's climate to inform our understanding of natural climate variability and future climate change (*see Box 6. Understanding Future Climate by Looking Back in Time*).

Modeling Global Change

Models are critical tools for enhancing scientific understanding, making predictions and projections, and ensuring informed decision making, and USGCRP agencies continue to make advancements in this important domain (*see Box 7. Combining Models to Better Predict Future Conditions*). USGCRP coordinates global change modeling efforts across the Federal Government and brings together researchers from different disciplines to develop models that integrate across diverse Earth system components. Atmospheric, oceanic, terrestrial, and cryospheric components of the Earth system have been part of integrated climate models for a number of years, and can now be represented with increased spatial resolution. More recently, other key components have been incorporated, such as ice sheets, aerosols, land hydrology, vegetation, and biogeochemical cycles. Additionally, socioeconomic drivers of global change are now beginning to be included more comprehensively and dynamically in models.

USGCRP continues to focus on deeper integration of physical climate models with other Earth system components and societal factors in order to improve the utility of models for decision makers. Recent integrated assessment modeling efforts, for instance, couple socioeconomic factors and natural processes (simulated by Earth system models) in a single interdependent framework, allowing researchers and decision makers to explore a range of global change scenarios, such as the impacts of water and land use on the climate system and the implications of water availability on human adaptation and energy choices. Under the President's Climate Action Plan, USGRCP is charged with establishing a

Box 4. Continuing Critical Earth Observations with Landsat Satellites

Overview

The Landsat Data Continuity Mission (LDCM), also known as Landsat 8, is the latest satellite in the joint NASA-U.S. Geological Survey (USGS; DOI) program that has been continuously tracking Earth surface changes like forest loss, glacial retreat, and urban buildup since 1972, when the program's first satellite was launched. Landsat 8, which was launched on February 11, 2013, continues the program's critical contribution to monitoring, understanding, and managing essential resources such as food, water, and forests, as well as the impacts of human society on the planet. Landsat 8 joined the aging Landsat 7 in orbit and is producing further stunning pictures of Earth's surface along with a wealth of scientific data.

Research and Societally Relevant Outcomes

Landsat 8 measures the Earth's surface in the visible, near-infrared, short-wave infrared, and thermal infrared spectra, with a resolution of 15 to 100 meters. The knowledge gained from 40 years of continuous Landsat observations contributes to research on climate, the carbon cycle, the water cycle, ecosystems, and natural and human changes to the Earth's surface. Through such research, the Landsat imaging time series has led to new results on the effects of land use change on human health and biodiversity, and has ultimately led to improvements in energy and water management, urban planning, disaster recovery, and agricultural monitoring.

Looking ahead, NASA will develop a long-term plan for sustained land imaging with support from USGS. NASA and USGS are currently evaluating sustainable, cost-effective approaches with input from stakeholders such as the research community, the industry sector, state and local governments, and international partners.

The following links provide additional information:
landsat.usgs.gov/
nasa.gov/mission_pages/landsat/main/

An artist's rendition of Landsat 8 monitoring from space. (Source: NASA)

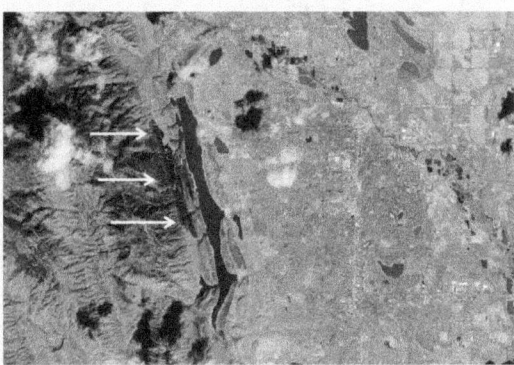

The first image from the Operational Land Imager (OLI) instrument aboard Landsat 8 shows the area around Ft. Collins, CO, at 15 m resolution using natural color. A charcoal-colored burn scar from the 2013 Galena Fire is indicated with arrows. (Source: NASA)

public-private partnership to explore modeling of risks associated with extreme weather events, including consideration of how the frequencies and intensities of these events will be affected by climate change. Such a public-private partnership may serve as a model for implementation of other decision support activities. At the international level, USGCRP supported the U.S.'s scientific contribution to the Fifth Assessment

Box 5. Supporting Global Change Research Through Satellite Missions: A Look Ahead

Overview

A series of Earth observation missions planned by NASA and partners for FY 2014 will contribute fundamentally to advancing our understanding of global change. Such missions are foundational to USGCRP research and are made possible by a sustained Program emphasis in instrumentation development. The planned FY 2014 missions are described below.

Research and Societally Relevant Outcomes

The Total solar irradiance Calibration Transfer Experiment (TCTE), funded by DOC's National Oceanic and Atmospheric Administration (NOAA) and launched aboard a DOD satellite, will use modified residual NASA hardware to mitigate a potential upcoming gap in an otherwise continuous 34–year record of total solar irradiance provided by previous NASA missions. Total solar irradiance is an important external driver of the climate system, and therefore crucial to a comprehensive understanding of climate change.

The Global Precipitation Measurement (GPM) mission, a joint effort with the Japan Aerospace Exploration Agency (JAXA), will provide the first opportunity to calibrate measurements of precipitation across tropical, mid-latitude, and polar regions. GPM will contribute to a better understanding of the global water and energy cycles; support climate predictions by providing foundational information about surface water fluxes, soil moisture storage, and other factors affecting the climate system; and improve capabilities for predicting natural hazards (e.g., floods, drought, landslides, hurricanes) and fresh water availability.

Preparing equipment for the Global Precipitation Measurement mission. (Source: NASA)

The Orbiting Carbon Observatory-2 (OCO-2) will measure the concentration of atmospheric CO_2 from space. Through high-resolution global coverage, OCO-2 will enable scientists to monitor the geographic distribution and variability of CO_2 sources (emissions) and sinks (uptake).

The ISS-RapidSCAT will be launched to the International Space Station (ISS), where it will measure ocean-surface wind speed and direction. These measurements will improve weather forecasts (including hurricane monitoring) and enable a better understanding of how ocean-atmosphere interactions influence Earth's climate. The ISS-RapidSCAT will serve as a calibration standard, helping scientists to merge ocean-surface wind data from multiple sources into a long-term, consistent record.

The Cloud Aerosol Transport System (CATS) will also be launched to the ISS, and will use advanced lidar technology to measure clouds and aerosols like pollution, dust, and smoke. The system will enable study of day-to-night changes, which Earth science satellites typically cannot offer because of their orbits. CATS will continue observations of Earth's changing atmosphere that help us to understand formative and ongoing processes, and ultimately model, predict, and plan for future climate changes.

The following links provide additional information:
TCTE: lasp.colorado.edu/home/missions-projects/quick-facts-tcte/
GPM: science.hq.nasa.gov/missions/earth.html
OCO-2: science.nasa.gov/missions/oco-2/
RapidSCAT: jpl.nasa.gov/missions/iss-rapidscat/
CATS: nasa.gov/mission_pages/station/research/experiments/1037.html

An artist's rendition of the Global Precipitation Measurement satellite in space. (Source: NASA)

Box 6. Understanding Future Climate by Looking Back in Time

Overview

A number of USGCRP agencies such as NSF, DOE, NOAA, NASA, USGS, and SI are investing in understanding the history of Earth's climate, known as "paleoclimate." Paleoclimate data extend records of climate and the environment beyond the time period for which we have instrumental measurements. Studying paleoclimate not only answers questions about what Earth was like in the past, but also provides critical context for the climate changes that we are experiencing today, and informs our understanding of how climate is likely to change in the future. A few examples of USGCRP agencies' investments in paleoclimate research and education are described below.

Research and Societally Relevant Outcomes

NSF and international partners from 12 countries support the North Greenland Eemian Ice Drilling (NEEM) project, which focuses on the Eemian interglacial period. The Eemian climate was similar to that of the Holocene (which includes the present day), but temperatures were approximately 3–5°C warmer than they are currently, making the Eemian a useful analog for expected future temperature increases of 2–4°C by 2100. The NEEM project drilled a core 2.5 kilometers deep into the Northwest Greenland ice sheet, capturing ice from the Eemian through the present day. Analyses of this ice core have resulted in 25 publications since 2012 alone, ranging from topics such as the isotopic record of carbon monoxide to likely patterns of future ice melt and resultant sea-level rise.

The Paleoclimate Modelling Intercomparison Project (PMIP) is an international effort that involves NOAA, DOE, NSF, and NASA. PMIP uses state-of-the-art climate models to simulate paleoclimate, and evaluates the capabilities of those models to reproduce climate conditions that are radically different from those of today. Such modeling efforts depend on archives of paleoclimate data derived from sources like tree rings, ice cores, corals, and ocean and lake sediments. NOAA's National Climate

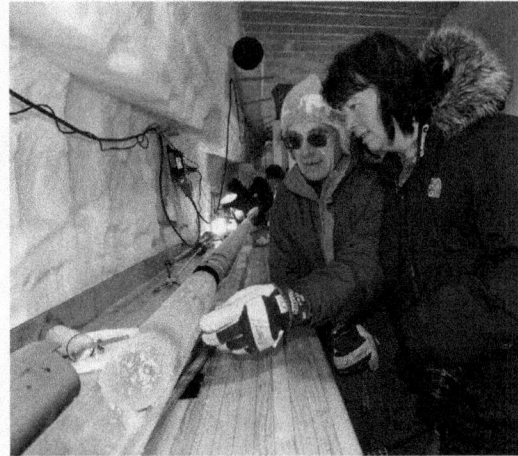

Scientists examining an ice core used to reconstruct the evolution of Earth's climate during the Eemian interglacial period, a useful analog for present-day and expected future conditions. (Source: NEEM)

Data Center archives extensive paleoclimate data that are contributed and used by scientists all over the world. Paleoclimate data are also used in Earth and environmental science education: SI has developed lesson plans for students at a range of grade levels that demonstrate how researchers use sources of information about the past—like leaf fossils, ancient coral reefs, and even archeological artifacts—to reconstruct past climate conditions.

The following links provide additional information:
NEEM: neem.dk/
PMIP: pmip.lsce.ipsl.fr/
National Climate Data Center (paleoclimate): ncdc.noaa.gov/data-access/paleoclimatology-data
SI education: smithsonianeducation.org/educators/lesson_plans/lesson_plans.html

Report of the Intergovernmental Panel on Climate Change (IPCC), including through the involvement of USGCRP agencies in Phase 5 of the Coupled Model Intercomparison Project (CMIP5) (*see Box 8. Modeling 20th and 21st Century Climate to Understand and Predict Change*). Going forward, USGCRP will develop descriptions of the computational and data infrastructure needed to conduct scientific research in support of future IPCC assessments.

Box 7. Combining Models to Better Predict Future Conditions

Overview

The North American Multi-Model Ensemble (NMME)—led by NOAA in partnership with DOE, NSF, NASA, and U.S. and Canadian research institutions—is an experimental climate forecasting system that combines a suite of different models. Using a combined system of models, each with different individual strengths in predicting phenomena, can enhance the overall success of a prediction beyond what would be possible with a single model.

The following link provides additional information:
cpc.ncep.noaa.gov/products/NMME/

Research and Societally Relevant Outcomes

Although NMME was developed as a research experiment, NOAA has successfully used the system to inform climate predictions. An operational version of NMME—where "operational" refers to use of the system as part of the national climate prediction capacity—is under development and shows promise. For example, model forecasts initialized in November 2011 were able to effectively predict temperatures for March 2012—four months later. NMME is thus expected to improve our ability to forecast climate conditions anywhere from months to seasons in advance.

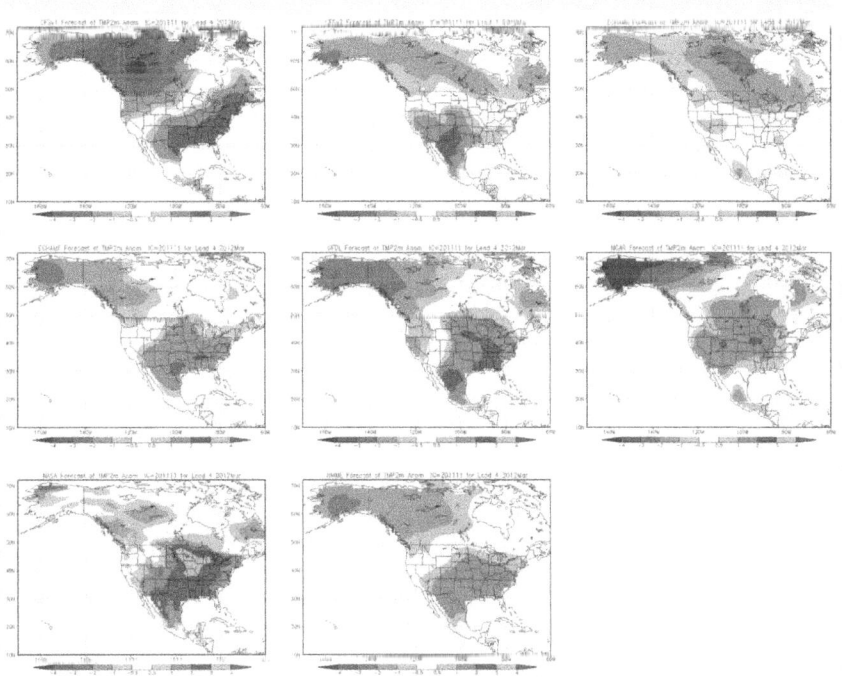

A suite of forecasts of air temperature ~6.5 feet (2 meters) above the ground, initialized in November 2011, were able to effectively predict March 2012 conditions. Red indicates higher temperatures and blue indicates lower temperatures relative to historical averages from 1982 to 2010. (Source: NMME)

Advancing Use-Inspired Science for Adaptation and Mitigation

USGCRP also works to advance use-inspired science driven by issues important to society, such as adapting to and mitigating climate risks. The Program provides the scientific foundation for risk management in the areas of greatest societal need, as defined and guided by the strategic priorities of decision makers and stakeholders. USGCRP engages with these stakeholders in a number of ways—through decision support and assessment activities, as well as communication and education efforts described in the 10-year Strategic Plan.

3.2 Preparing for the Impacts of Climate and Global Change

Across the Nation and around the world, people are making decisions to build resilience to, prepare for, and minimize the impacts of global change. Over more than two

decades, USGCRP and its member agencies have built a strong scientific foundation to better understand potential impacts, risks, vulnerabilities, opportunities, and trade-offs related to climate change. In FY 2013, USGCRP strengthened this scientific foundation to support practical, on-the-ground decisions by conducting domestic and international assessments, sharing data to inform decision makers, and educating, engaging, and communicating with the Nation about climate-change impacts, preparedness, and adaptation efforts.

Box 8. Modeling 20th and 21st Century Climate to Understand and Predict Change

Overview
Climate and Earth system modeling supported by USGCRP agencies such as DOE, NASA, NOAA, NSF, and others was foundational to the IPCC's Fifth Assessment Report (AR5) on the scientific basis for climate-change, released in 2013. Along with international partners, Federal and Federally supported modeling centers contributed to Phase 5 of the Coupled Model Intercomparison Project (CMIP5), a critically important information source for AR5 that focused on simulations of 20th and 21st century climate.

Research and Societally Relevant Outcomes
Modeling efforts like CMIP5 significantly advance our ability to predict (and thereby to plan for) changing climate conditions. Simulations of 20th century climate allow scientists to

evaluate the accuracy of models by comparing results with climate observations, such as those supported by USGCRP. Simulations of 21st century climate form the foundation of climate change projections. CMIP5's 21st century simulations were driven by greenhouse gas emissions scenarios produced through integrated assessment modeling (also supported by USGCRP agencies), which couples natural processes with an extensive set of infrastructure, land use, and socioeconomic information.

Scientists affiliated with and supported by USGCRP agencies analyzed the data produced by CMIP5, leading to the publication of articles that are extensively cited in scientific journals and the IPCC report. USGCRP agencies also held a workshop on the science and process of CMIP5 to inform future phases of the project. In addition to key presentations by researchers supported by DOD, DOE, NASA, NOAA, NSF, and USDA, the workshop included presentations from the NOAA CMIP5 Task Force, which organized a special collection of scientific papers in the Journal of Climate covering 20th and 21st century simulations of North American climate.

The following links provide additional information:
cmip-pcmdi.llnl.gov/cmip5/
cpo.noaa.gov/MAPP/CMIP5TF

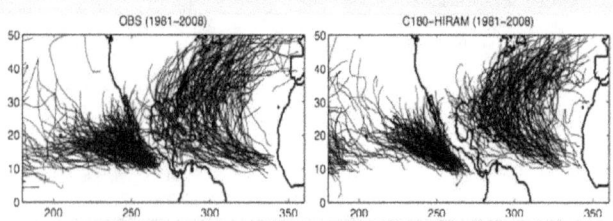

A comparison of observed (left) and simulated (right; single model realizations) hurricane tracks around North America between 1981 and 2008 (the horizontal and vertical axes show longitude and latitude, respectively). Tropical cyclones are challenging to simulate, but these CMIP5 results demonstrate promise that such storms can be effectively represented in models. (Source: Adapted from Sheffield et al., 2013[1])

[1] Sheffield, J. et al., 2013. North American Climate in CMIP5 Experiments. Part II: Evaluation of Historical Simulations of Intraseasonal to Decadal Variability. J. Climate, 26, 9247–9290. doi: dx.doi.org/10.1175/JCLI-D-12-00593.1 © American Meteorological Society.

Assessments help scientists and decision makers anticipate likely changes, evaluate scientific information for
potential use in decision making, and pinpoint knowledge gaps and needs. Here, "assessments" refers to
syntheses of peer-reviewed literature and other credible
sources that convey the current scientific understanding
of global change. Some assessments focus on specific
geographical regions, while others focus on specific
aspects or impacts of global change.

Since its inception, USGCRP has placed significant
emphasis on national and international assessments.
USGCRP is required to conduct a National Climate
Assessment (NCA) every four years and to coordinate
Federal participation in international assessment efforts
such as those led by the IPCC. Recently, both the NCA
and the IPCC have been improving the utility, accessibility, and transparency of assessment data by providing
digital access to information.

In FY 2014, USGCRP delivered its Congressionally
mandated quadrennial NCA report, in the form of
a comprehensive synthesis of climate science and
climate-change impacts on regions and sectors of the
United States. The NCA provides an important basis
for developing specific climate-relevant scenarios that
are useful for scientists and decision makers—including those that incorporate complex variables, such as
changing patterns of extreme events at regional scales.
In addition, USGCRP and its NCA program are helping
to coordinate an interagency effort to develop a set of
national climate indicators, which are physical, ecological, and societal variables that communicate key aspects
of how the climate is changing (*see Box 9. Developing
Indicators of Climate Change*).

A new area of focus for USGCRP is strengthening the
Program's capacity to conduct assessments on an it-
erative and sustained basis, rather than only at regular
multi-year intervals. USGCRP is developing a sustained
assessment process that will ultimately enable continuous
and transparent participation of scientists and stakeholders across regions and sectors, enabling new information and insights to be synthesized as they emerge.

The long-term strength of U.S. global change research
depends on close engagement with international efforts, including international climate assessments, and
USGCRP continues to ensure that U.S. interests are
represented internationally. For example, the Program
coordinates and supports U.S. participation in and review of the IPCC's Assessment Reports and the World
Meteorological Organisation/U.N. Environmental
Programme's (WMO/UNEP's) Scientific Assessment
of Ozone Depletion. Scientists affiliated with and
supported by USGCRP agencies continue to play key
roles in the development of such major international
assessments: they are lead authors, working group co-
chairs, and reviewers who provide technical support
and scientific expertise.

The IPCC has released the first three volumes of its
Fifth Assessment Report (AR5), and will release the
final Synthesis Volume and begin planning for subse-
quent assessments this year. USGCRP is coordinating
and supporting the U.S. Government's review, revi-
sion, and vetting of AR5. USGCRP will also engage in
the development of the 2014 WMO/UNEP Scientific
Assessment of Ozone Depletion, the content of which
relates closely to climate and global change.

*Understanding Global Change Preparedness,
Adaptation, and Resilience*

USGCRP agencies are working diligently to increase
dialogue between climate researchers and decision
makers to ensure mutual understanding of the most
pressing scientific and decisional needs. They are
also leveraging internal expertise to prepare for the
impacts of climate change. For example, NASA is
sponsoring focused research on potential approaches
to climate adaptation in the regions where its facilities
are located. NSF's four Centers for Decision Making
Under Uncertainty are continuing their work on how
effective decisions can be made across a number of
environmental policy issues when long-term outcomes
related to climate change are uncertain. Other agen-

cies are collaborating with state and local officials to co-produce informational products, tools, and resources relevant to managing extreme weather events, climate variability, and future change at local, state, and regional levels. Following Hurricane Sandy in October 2012, for instance, Federal agencies partnered with local decision makers to develop a set of informational products that illustrate both current and future risks

---

Box 9. Developing Indicators of Climate Change

Overview

Indicators are measurements or calculations that represent the status, trend, or performance of a system (e.g., the economy, agriculture, air quality). USGCRP, with the participation of 9 of its 13 member agencies—NOAA, NASA, EPA, USDA, DOE, DOD's U.S. Army Corps of Engineers (USACE), HHS's Centers for Disease Control (CDC), DOI, and NSF—is leading an interagency effort to identify a system of physical, ecological, and societal indicators that will inform and support decision making about climate changes, impacts, vulnerabilities, and responses.

This system of indicators is part of the vision of the sustained National Climate Assessment process (described in this section). It will include both current indicators and leading indicators (representing potential future states of the system), will be scalable for multiple levels of use, and will augment existing efforts where possible. For example, the system will leverage EPA's 2012 Climate Change Indicators Report, developed in partnership with several other Federal agencies and organizations. Although the 26 indicators identified in this report are not an exhaustive group of all climate indicators, they are a select set of key measurements relating to greenhouse gases, weather and climate, oceans, snow and ice, ecosystems, and human systems. These indicators are based on peer-reviewed data from government agencies, academic institutions, and other organizations.

Research and Societally Relevant Outcomes

The system of indicators is currently under development with the involvement of over 150 physical, natural, and social scientists and managers making up 13 thematic teams. Each team is charged with developing a conceptual model and providing topical recommendations intended to lead to the release of a limited set of indicators on the Global Change Information System (GCIS, see page 16) in 2014, followed by the launch of a broader set of indicators in 2015. In support of this effort, NASA has invested in

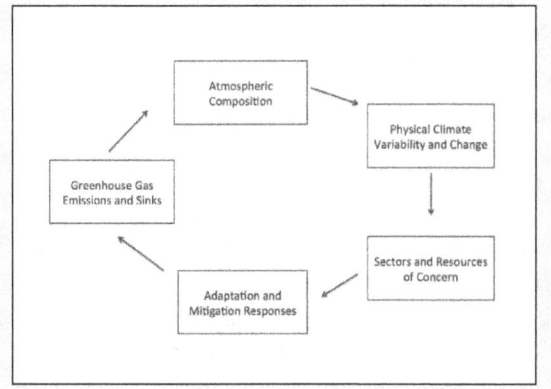

This conceptual framework shows linkages between categories of indicators.
(Source: USGCRP indicators program)

14 research projects to develop and test new indicators of climate change that could be incorporated into future assessment efforts.

Climate change indicators have the primary goal of informing users about climate in the broader, interconnected context of global change, as well as the far-reaching impacts of climate change on the environment and society. The outcomes and products of this effort are designed to address questions important to multiple audiences—including scientists, analysts, managers, planners, policy makers, educators, and the public—in a conceptually unified framework.

The following links provide additional information:
www.globalchange.gov/browse/reports/national-climate-assessment-physical-climate-indicators-workshop-report
epa.gov/climatechange/science/indicators/
weather.msfc.nasa.gov/nca/NASA_NCA_Indicators_Solicitation_-_Selected_Proposals.html

associated with sea-level rise and coastal inundation (*see Box 10. Planning for the Future After Hurricane Sandy*). These tools are intended to help communities plan for the future and build resilience through infra-

structure and construction efforts. Taking climate-change impacts like sea-level rise into consideration when rebuilding coastal areas can reduce the cost of future disasters and potentially save lives.

Box 10. Planning for the Future After Hurricane Sandy

Overview

Hurricane Sandy hit the northeastern United States in October 2012 and was the deadliest hurricane of the season, as well as the second costliest hurricane in U.S. history. Such extreme coastal flooding events are expected to become more frequent as a result of climate change-related sea-level rise. A Hurricane Sandy Rebuilding Task Force[1] was created to provide consistent, clear, accessible information for decision makers involved in recovery and rebuilding efforts.

As part of this effort, an ad-hoc working group of several Federal entities including the Federal Emergency Management Agency (FEMA), NOAA, USACE, the Council on Environmental Quality (CEQ), and USGCRP, in coordination with local institutions in New York City, developed a set of products that illustrate current and future risks of sea-level rise and coastal inundation.

Research and Societally Relevant Outcomes

Two major products are associated with this interagency effort. First, NOAA led the development of a set of map services that integrate the best available FEMA flood hazard data with scenarios of future sea-level rise for 2050 and 2100, allowing communities, residents, and other stakeholders to consider risks from sea-level rise in planning for reconstruction following Hurricane Sandy. Second, in partnership with FEMA and NOAA, USACE modified its existing Sea-Level Change Calculator to integrate FEMA's best available elevation data with the same sea-level rise scenarios. The USACE calculator complements the NOAA maps by providing site-specific detail on projected flood elevations for five-year intervals from 2010 to 2100. In conjunction with other local data, such information can be used by floodplain managers, professional engineers, and surveyors for developing additional safety margins above FEMA's best available elevation data.

In the aftermath of Hurricane Sandy, these products provide information that planners and decision makers need to increase resilience to future extreme events. Coastal commu-

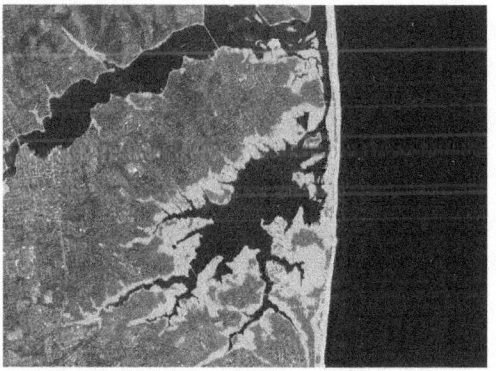

Map of current (yellow) and potential future (pink) flood risks for a portion of the New Jersey coastline. The different shades of pink represent four future scenarios with sea-level rise ranging from 8 inches to 6.6 feet. (Source: Sea Level Rise Map Services produced by NOAA in partnership with FEMA, USACE, CEQ, and USGCRP)

nities are likely to see further destructive flooding as a result of sea-level rise, but accounting for climate-change impacts when rebuilding coastal areas can reduce the cost of future disasters and potentially save lives.

The following links provides additional information:
www.globalchange.gov/browse/sea-level-rise-tool-sandy-recovery

---

1 The Hurricane Sandy Rebuilding Task Force includes the following Federal departments and other entities: Housing and Urban Development, Treasury, DOI, USDA, DOC, Labor, HHS, DOT, DOE, Education, Veterans Affairs, Homeland Security, EPA, Small Business Administration, USACE, Office of Management and Budget, National Security Staff, Domestic Policy Council, National Economic Council, CEQ, Office of Science and Technology Policy, Council of Economic Advisors, Office of Public Engagement and Intergovernmental Affairs, and Office of Cabinet Affairs.

In addition to supporting communities across the United States with cutting-edge global change science and research, USGCRP and its member agencies are working closely with other partners and interagency groups to provide actionable science in support of regional decision making. Because most adaptation and many mitigation decisions are made at local, state, and regional scales, rather than at the national level, deploying science and services that are tailored to regional decision making needs is key. Going forward, USGCRP agencies will work with partners to expand regional capacity to support decisions with scientific information; contribute to the developing sustained assessment process through support from NOAA's Regional Integrated Sciences and Assessments (RISA) program and other NCA partners; and support regional Landscape Conservation Cooperatives (coordinated by the U.S. Fish and Wildlife Service) and the DOI Climate Science Center network.

Sharing and Managing Critical Global Change Data and Information

Informed decisions require credible information and data. Decision makers need global change information that is centrally accessible, clearly described, authoritative, and relevant. Meeting this need efficiently—when both the demand for and the diversity of available data are expanding rapidly—is a significant challenge. To address this, USGCRP is developing a new, systematic approach to global change information provision. The approach confronts the reality that while many credible, topic-specific data delivery services exist across the Federal Government, there is no single point of access for authoritative information on interrelated, multidisciplinary global change issues such as the coastal impacts of sea-level rise, the health costs associated with temperature extremes, and others. Without a central and intuitively structured access point, it can be difficult for users to find the data they need.

The main component of USGCRP's information provision plan is the implementation of a Global Change Information System (GCIS). The GCIS—currently under development—is intended to be a comprehensive and integrative web-based platform that will efficiently deploy the broad range of global change information and data to diverse user communities. The GCIS will leverage and build upon existing capabilities in information technology and data systems from NASA, NOAA, USGS, and other agencies to transparently link agency data sources. Using semantic web approaches, the GCIS will provide traceability between multiple environmental data streams (e.g., data from sensors) and the resulting models, publications, and reports. Additionally, for data associated with USGCRP, the GCIS will serve as a mechanism for tracking usage and identifying users, which is important for USGCRP performance metrics. As a first but significant step, the GCIS provides traceability for and access to information and data related to the Third NCA report. Subsequently, the GCIS will expand to include a wider portfolio, with an initial emphasis on climate-related information that is relevant to human health and adaptation planning.

Engaging, Educating, and Communicating with Stakeholders

As the leading Federal authority on global change science, USGCRP and its member agencies are well positioned to engage and educate citizens about global change and related societal issues. USGCRP helps to coordinate the development of multiagency products and programs, grow and expand the reach of information beyond single agencies, and ensure that feedback from public engagement is shared broadly within the Federal global change science community. In addition, USGCRP develops a national climate change education communication strategy that includes all member agencies, and coordinates climate education, communication, and engagement priorities within the Program. USGCRP agencies also play leadership roles in developing the scientific workforce of the future. As one example, NSF, NOAA, and NASA have partnered with the American Meteorological Society (AMS) for the AMS Climate Studies Diversity Project, which develops capacity among faculty at minority serving institutions (MSIs) to implement climate-related coursework, thereby enhancing climate science learning opportunities for both educators and students (*see Box 11: Building Capacity and Diversity in Climate Studies*).

Box 11. Building Capacity and Diversity in Climate Studies

Overview

The American Meteorological Society (AMS) Climate Studies Diversity Project, funded primarily by NSF, leverages resources from NOAA, NASA, and academic and non-profit partners to build climate education capacity at minority serving institutions (MSIs). Through a five-day professional development workshop, the project trains MSI faculty to implement the introductory-level AMS Climate Studies course at their institutions. The workshop is designed for faculty both with and without previous Earth science teaching experience. Participants are immersed in the course materials and, by the end of the workshop, demonstrate the ability to interpret and analyze climate data acquired through direct observations and remote sensing; an understanding of climate from a dynamic, Earth system perspective; and an understanding of the course delivery system and a variety of implementation strategies. Workshops began in 2012 and are held annually, reaching ~25 faculty members each round.

Research and Societally Relevant Outcomes

The United States faces a serious challenge in attracting young people, and especially members of underrepresented groups, to Earth science careers (including teaching), in part because of limited opportunities to enroll in introductory-level courses. The AMS Climate Studies Diversity Project helps to address this problem by enhancing climate science learning opportunities for both educators and students, and has already led to the implementation of the AMS Climate Studies course at 52 MSIs.

Engagement with the project continues after the workshop and extends beyond the implementation of the course: par-

Participants in the AMS Climate Studies Diversity Project workshop in Washington, DC.
(Source: AMS)

ticipants share best-practice strategies learned from teaching the AMS Climate Studies course with their professional community, and engage in a network connecting students with scientists for internships, research, and career counseling. Participants also receive support to attend the AMS Annual Meeting, where they present their progress at the Education Symposium.

The following link provides more information:
ametsoc.org/amsedu/online/climateinfo/diversity.html

USGCRP also supports education and capacity building efforts at the international level. For instance, USGCRP provides funding for START (the Global Change SysTem for Analysis, Research and Training), an internationally recognized, U.S.-based non-profit that works to strengthen capacity in developing countries for understanding and addressing challenges associated with global change. START's African Climate Change Fellowship Program (ACCFP), for example, helps develop endogenous capacity for advancing and applying climate adaptation strategies in Africa (*see Box 12. Developing International Capacity for Climate*

Adaptation). In addition, START's USAID-funded Cities at Risk program supports the integration of scientific information about climate-change impacts, vulnerabilities, and adaptation into urban planning and policy in Asian and African coastal cities. Enabled by annual support from USGCRP, START's portfolio of capacity building programs also includes curricula development at African universities, communication trainings to promote science for decision making, and national- and regional-scale dialogues that create opportunities for exchange among scientists, policymakers, practitioners, and the media.

Box 12. Developing International Capacity for Climate Adaptation

Overview

Through its annual funding of START (the Global Change SysTem for Analysis, Research and Training), USGCRP supports the African Climate Change Fellowship Program (ACCFP), an effort that is strengthening a growing pan-African knowledge network on adaptation to climate change. The ACCFP was pioneered in 2007 by START in partnership with the University of Dar Es Salaam, the African Academy of Sciences, and the Climate Change Adaptation in Africa program.

ACCFP fellows are matched with universities, research centers, and other host institutions across Africa where they work with mentors to implement individually designed projects. Supported by access to USGCRP agencies' climate and remote sensing data, as well as collaborating U.S. scientists, these projects assess and prioritize climate risks, consider approaches for integrating adaptation with planning and practice, and investigate current practices for designing and implementing adaptation actions. Having engaged more than 120 fellows and 100 institutions to date, the ACCFP is recognized as a major platform for education, training, and capacity building in Africa.

Research and Societally Relevant Outcomes

Creating endogenous capacity for climate adaptation is important throughout the world. It is especially critical in Africa, which is highly vulnerable to the impacts of climate change, largely because of the convergence of several factors that amplify climate risks. These include endemic poverty, high dependence on natural resources, poor access to basic necessities, inadequate infrastructure, and challenges of

Farmers in Kenya meet to discuss agricultural risk management. (Source: START)

governance. The individual ACCFP projects, each of which is rooted in particular places, communities, and research questions, produce results that together provide a narrative of climate change and adaptation across Africa, informing current understanding and decision making as well as future priorities for program design and development.

In 2012, USGCRP co-funded an ACCFP Writing Retreat that led to the publication in 2013 of an open-access special issue of the journal *Environmental Development*, which focused on climate change risk management in Africa.

The following links provide additional information:
start.org/programs/accfp1
sciencedirect.com/science/journal/22114645/5

4 LEVERAGING GLOBAL CHANGE RESEARCH IN AREAS OF SOCIETAL NEED

USGCRP's research investments are being used today to better inform responses to global change in high-priority areas of national need, including: (1) extreme weather and climate events; (2) sea-level rise and melting ice; and (3) climate change and human health. The following are illustrative examples of USGCRP research investments in these three cross-cutting areas.

4.1 Extreme Weather and Climate Events

Across the Nation, Americans are noticing important climatic changes, including hotter summers, longer periods of extreme heat, heavier downpours, and longer dry spells in certain parts of the country. These changes, which vary by region, are having important impacts on local businesses, communities, and families. Observed changes in climate extremes reflect the influence of anthropogenic climate change in addition to natural climate variability. Observations also show that changes in exposure and vulnerability to extreme events are influenced by both climatic and non-climatic factors.[4] Effective sharing of resources, information, and data can improve the understanding of and response to extreme weather events. For example, USDA and NOAA recently entered into a memorandum of understanding (MOU) to improve sharing of data and expertise, monitoring networks, and forecasting efforts related to drought (*see Box 13. Sharing Information to Improve Drought Monitoring and Forecasting*).

Other examples of effective interagency collaboration in this domain include partnerships among NOAA, USGS, and USACE to improve forecasting capabilities for some of the most destructive climate-related disasters: storms (including hurricanes), droughts, and floods (*see Box 14. Forecasting Extreme Events to Increase Preparedness and Resilience*). In addition, Federal entities—including FEMA, NOAA, USACE, CEQ, and USGCRP—partnered with local decision makers after Hurricane Sandy to develop a set of informational products that illustrate

both current and future risk from coastal inundation and sea-level rise (*recall Box 10. Planning for the Future After Hurricane Sandy*).

4.2 Sea-Level Rise and Melting Ice

Like mercury in a thermometer, water expands as it warms up ("thermal expansion"), causing sea levels to rise as the climate warms. Melting of glaciers and ice sheets is also contributing to sea-level rise at increasing rates. Since 1900, tide gauges throughout the world have shown that global sea level has risen by about 8 inches; since 1992, the rate of global sea-level rise measured by satellites has been nearly twice the rate observed over the last century. Sea level on the U.S. Atlantic Coast is rising faster now than at any time in at least the past 2000 years. Sea-level rise is expected to continue beyond the end of this century, with significant implications for coastal communities, populations, and businesses in the United States.

Projecting future rates of sea-level rise is challenging. Even the most sophisticated climate models cannot precisely simulate recent rapid changes in ice sheet dynamics. In recent years, "semi-empirical" models have been developed, which are based on statistical relationships between historical rates of global warming and sea-level rise. In December 2012, NOAA released Global Sea Level Rise Scenarios for the United States National Climate Assessment,[5] reporting high confidence among scientists that global mean sea level will rise at least another 8 inches (0.2 meters) and no more than 6.6 feet (2.0 meters) by

---

[4] IPCC, 2012. Summary for Policymakers. In: Managing the Risks of Extreme Events and Disasters to Advance Climate Change Adaptation. A Special Report of Working Groups I and II of the Intergovernmental Panel on Climate Change. Website: ipcc-wg2.gov/SREX/images/uploads/SREX-SPMbrochure_FINAL.pdf

[5] Website: scenarios.globalchange.gov/report/global-sea-level-rise-scenarios-united-states-national-climate-assessment

2100. This important report was produced at the request of a Federal advisory committee charged with developing the next National Climate Assessment (NCA) and includes input from national experts in climate science, physical coastal processes, and coastal management. The report provides a synthesis of the scientific literature on global sea-level rise and a set of four future scenarios to help assess potential vulnerabilities and impacts.

Over time, gradual sea-level rise and more frequent extreme weather events (such as hurricanes) can alter the coastline. Federal agencies are partnering with communities to understand recent trends in regional shoreline change along U.S. coasts. Several USGCRP agencies including USGS, USDA, NOAA, and USACE, along with other partners, began the National Assessment of Shoreline Change Project (NASCP) to establish a set of

Box 13. Sharing Information to Improve Drought Monitoring and Forecasting

Overview
The 2011–2012 U.S. drought was the most severe and extensive in at least 25 years, and the effects are still being felt in some areas of the Nation. Recently, USDA partnered with local governments, colleges, and state and Federal partners—including NOAA and FEMA—to conduct a series of regional drought workshops. Hundreds of farmers, ranchers, business owners, and other stakeholders met with government officials to discuss needs and available programs. Agriculture Secretary Tom Vilsack kicked off the first meeting in Nebraska, and additional meetings were held in Colorado, Arkansas, and Ohio.

Research and Societally Relevant Outcomes
As a direct outcome of this series of workshops, USDA and NOAA entered into a memorandum of understanding (MOU) to improve sharing and coordination of drought-related data and expertise, monitoring networks, and forecasting efforts. The MOU is an effort to meet the needs identified by stakeholders at the workshops.

The frequency and intensity of droughts in the U.S. is likely to increase as climate changes. Agriculture is the largest sector of the U.S. economy impacted by drought and heat waves, but many others can be affected, such as tourism, energy, transportation, and industries that rely on high water levels to ship products. Improved sharing of data and expertise between agencies will lead to more coordinated, accurate efforts in drought monitoring and forecasting. NOAA and DOE have also established an MOU to coordinate research involving future climate extremes.

The following link provides additional information: blogs.usda.gov/2012/12/13/usda-national-oceanic-and-atmospheric-administration-enter-into-agreement-to-improve-drought-weather-forecasting/

USDA Agriculture Secretary Tom Vilsack speaks in Omaha, NE at the opening session of a drought workshop with stakeholders. (Source: USDA)

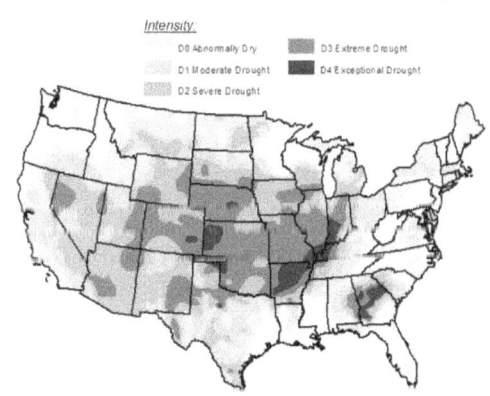

A map of continental U.S. drought conditions during July 2012, showing extreme and exceptional drought in several regions. (Source: U.S. Drought Monitor, a partnership between USDA, NOAA, and the National Drought Mitigation Center)

modern baseline data. A primary goal of this work is to develop standardized methods for mapping and analyzing shoreline movement that will ensure internal consistency (*see Box 15. Documenting Shoreline Changes to Better Understand Climate Impacts*).

Though summer sea ice in the Arctic Ocean has fluctuated in recent years, it has declined rapidly overall during the past several decades, with consequences for climate, coastal communities, and marine ecosystems. Ice loss means that highly reflective ice cover

Box 14. Forecasting Extreme Events to Increase Preparedness and Resilience

Overview

NOAA, USGS, and USACE are working together to help the country prepare for and manage the impacts of changing weather and climate patterns, particularly with regard to extreme events like hurricanes, floods, and drought.

Before, during, and after coastal storm events, USGS assesses the likelihood of beach erosion, overwash, or inundation, and provides information to NOAA on associated coastal vulnerability and change. USGS also measures storm surge and conducts real-time monitoring of inland rivers and streams. NOAA's National Weather Service relies on timely and accurate USGS data to issue flood warnings. Together, USGS, NOAA's National Weather Service, and USACE are developing flood inundation maps that show exactly where flood waters will be, street by street and hour by hour.

Research and Societally Relevant Outcomes

Interagency cooperation leads to better monitoring, understanding, and prediction of weather patterns and extreme events related to climate change. More accurate and timely weather warnings help to keep communities around the country safe, and emergency planning tools like the flood inundation maps can increase awareness of potential risks, contribute to preparedness and mitigation efforts, reduce economic losses, and even save lives.

The following links provide additional information:
water.usgs.gov/osw/flood_inundation/
coastal.er.usgs.gov/hurricanes/
waterwatch.usgs.gov

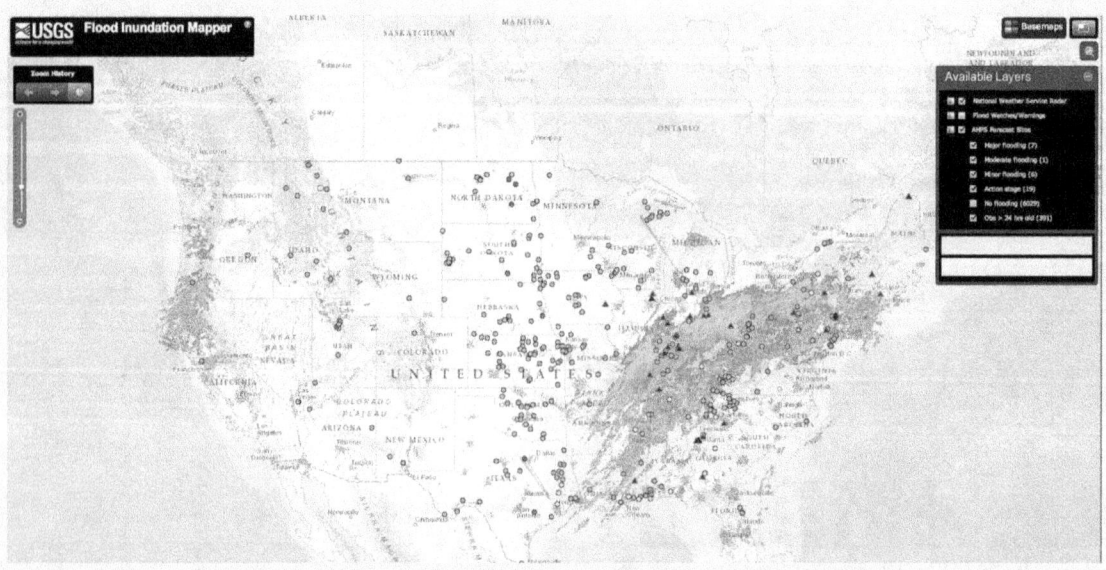

USGS, USACE, and NOAA's National Weather Service collaborated to create this Flood Inundation Mapper. (Source: USGS)

is replaced with open water, which absorbs more heat, contributing to a faster rise in temperatures in the Arctic than anywhere else on the planet. This increased heating causes additional ice melt, accelerating the cycle of sea ice decline in what is known as a "feedback loop." In

addition, sea ice serves as an important barrier to storm surge, meaning that sea ice loss can increase coastal erosion and vulnerability to storms. Changes in sea ice extent disrupt marine food webs, affecting fisheries and threatening the economic base and viability of coastal

Box 15. Documenting Shoreline Changes to Better Understand Climate Impacts

Overview
Coastal erosion is a long-term concern along most open-ocean shorelines in America. As coastal populations increase and more infrastructure is built to support them, demand is increasing for accurate information and regionally com-prehensive analyses regarding past and present shoreline changes.

In an effort to document and understand recent trends in regional shoreline dynamics, several USGCRP agencies including USGS, USDA, NOAA, and USACE, with help from others, began the National Assessment of Shoreline Change Project (NASCP) to establish a set of modern baseline data. A primary goal of this work is to develop standardized methods for mapping and analyzing shoreline movement, thereby ensuring internal consistency in records of erosion and accretion. In addition, the USGS Coastal and Marine Ge-ology Program's National Assessment of Shoreline Change Web Mapping Application—which includes data compiled in

support of NASCP—provides a map view of historical and modern shorelines, as well as short- and long-term shoreline change evaluations.

Research and Societally Relevant Outcomes
As sea level rises and severe storms like Hurricane Sandy become more common, understanding the processes under-lying coastal change will be more important than ever. One of NASCP's focus areas is the Arctic coast of Alaska, where scientists are documenting shoreline changes to better understand the geologic, oceanographic, and cryogenic pro-cesses that drive them. Alaska's Arctic coast is of particular interest because it is one of the most vulnerable coastlines in the United States; moreover, with increased thawing of Arc-tic summer ice, this coastline is becoming more important to U.S. national interests in security and natural resources.

The following link provides additional information: coastal. er.usgs.gov/shoreline-change/

The USGS's National Assessment of Shoreline Change Web Mapping Application shows modern and historical shorelines and evaluations of shoreline change. This screenshot shows long-term shoreline change data for a portion of the mid-Atlantic coast. (Source: USGS)

Arctic communities. At the same time, ice retreat could open up new regions for oil and gas exploration, and reductions in seasonal sea ice cover and warmer surface temperatures may create new habitat for some important fish species, such as cod, herring, and pollock.

USGCRP agencies are supporting a range of observations, process studies, and modeling efforts to understand changes in Arctic sea ice and to project likely changes over the next several decades. Detailed sea ice observation campaigns using lidar, radar, and other sensors are providing valuable calibration and validation for ongoing satellite observations of changes in sea ice extent. On site in the Arctic, US-GCRP agencies are using a variety of instruments and platforms, including unmanned aircraft, to measure ocean surface, subsurface, and atmospheric conditions in an effort to better understand the dramatic changes in sea ice over recent years.

Modeling studies range from investigations into why current models have underestimated the observed sea ice loss to projections of how changing sea ice distribution will affect potential shipping routes across the Arctic Sea, which has significant economic, strategic, environmental, and governance implications for the region. In addition, the Interagency Arctic Research Policy Committee (IARPC)—led by NSF in partnership with NOAA, DOD, DOS, HHS, USDA, DOE, DOI, DOT, NASA, EPA, SI, the Department of Homeland Security, the Office of Science and Technology Policy, and the National Endowment for the Humanities—is pursuing projects that will yield substantial advances in sea ice predictions at a variety of time and space scales through improvement of model physics. IARPC recently initiated the Sea Ice Prediction Network, which brings together scientists and stakeholders to improve and communicate sea ice prediction capabilities and results.

4.3 Climate Change and Human Health

Climate change poses significant threats to human health. Increasingly severe heat waves, for example, have already wrought serious negative impacts on

the elderly and other sensitive populations in parts of the United States and other regions of the world. Asthma rates have doubled in the past 30 years, and asthma attacks will increase as air pollution gets worse.[6] In addition, climate-related changes to oceans and coasts can impact health and safety, including through increased flooding (which can inundate sewers and landfills, leading to contamination of coastal and drinking waters), changes in harmful algal blooms, and diminished ability of ecosystems to provide services on which people depend. Climate change impacts such as increased temperature, sea-level rise, droughts, floods, and ocean acidification can also affect water and food supplies, as well as the occurrence of vector- and water-borne diseases. *Vibrio*, for example, is a type of bacteria that lives in warm coastal areas and is the most common cause of seafood-borne disease. Several USGCRP agencies and partners are working to reduce the prevalence of health risks associated with *Vibrio* bacteria (*see Box 16. A Global Initiative for Vibrio Early Warning Systems*).

The public health community has long recognized that environmental factors, including climate change, can have both direct and indirect impacts on human health. As climate change continues to boost the frequency and intensity of certain kinds of extreme weather events, and as temperature shifts and related phenomena alter the endemic ranges of disease-carrying mosquitoes and other vectors, it is crucial to understand how these phenomena could affect—and in some cases already are affecting—the occurrence and severity of health issues worldwide. To inform more effective responses to climate-related health threats, USGCRP—through a coordinated effort involving NOAA, NIH, CDC, NASA, EPA, and USGS—launched the Metadata Access Tool for Climate and Health (MATCH)[7] in May 2013. MATCH is a publicly accessible, searchable online clearinghouse of metadata (i.e., standardized informa-

---

6 The President's Climate Action Plan, 2013. Website: whitehouse.gov/sites/default/files/image/president27sclimateactionplan.pdf

7 Website: match.globalchange.gov

tion that describes and contextualizes data) extracted from more than 9,000 Federal datasets spanning public health, climate, oceans, and the environment. MATCH, which will ultimately be linked to the GCIS, will help researchers, health practitioners, and the public to understand, mitigate, and adapt to the health-related effects of climate change, while fostering collaboration among traditional and non-traditional partners.

Box 16. A Global Initiative for *Vibrio* Early Warning Systems

Overview

Vibrio is a type of disease-causing bacteria that occurs naturally in warm coastal areas. It is the most common cause of seafood-borne disease and is associated with 95% of related fatalities. To reduce the prevalence of *Vibrio*-related health risks in the United States and beyond, several USGCRP agencies including NOAA, NASA, NSF, and HHS's National Institutes of Health (NIH), along with other partners, began a Global Initiative for *Vibrio* Early Warning Systems. The goal of these systems is to build capacity to predict and reduce public health impacts from *Vibrio*, enhance existing data collection and integration, establish baseline health and environmental data, and facilitate adaptation to climate change and connection to climate services.

Research and Societally Relevant Outcomes

The prevalence of *Vibrio* is expected to increase as oceans become warmer in response to climate change. This initiative aims to determine key indicators that can be integrated to develop a predictive model for *Vibrio*-related outbreaks, with a particular emphasis on the species of *Vibrio* that causes cholera. Improved data and predictions for *Vibrio*-related outbreaks may help to prevent such outbreaks in the future.

The following link provides additional information:
oceansandhumanhealth.noaa.gov/

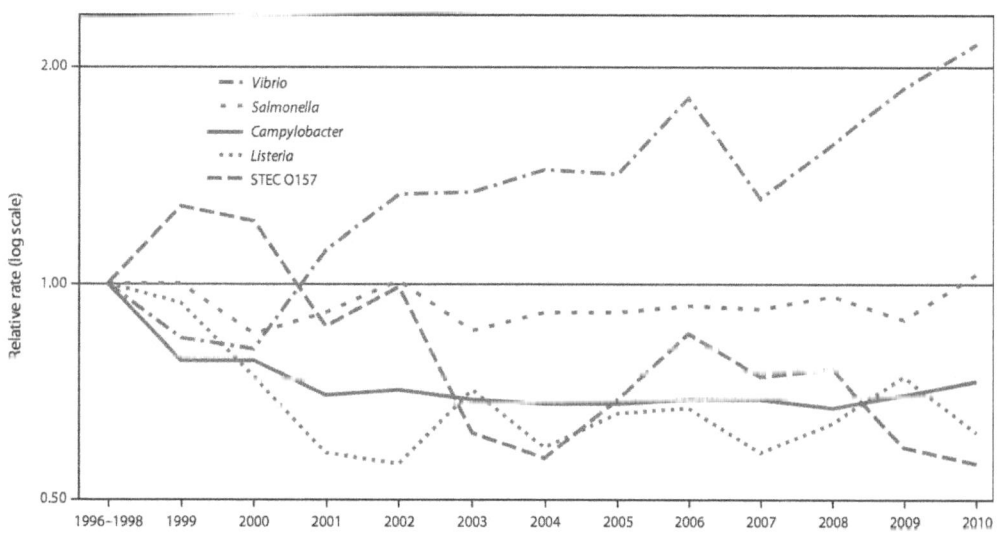

While the rates of infection for most food-borne diseases have decreased in recent years, the rate of *Vibrio* infections has increased (rates are shown relative to 1996–1998 rates, signified by the line intersecting 1.00 on the vertical axis). Because *Vibrio* thrives in warm waters, infection rates have the potential to rise as ocean temperatures increase. (Source: CDC)

5 USGCRP INTERAGENCY RESEARCH PRIORITIES FOR FY 2014

In accordance with the National Global Change Research Plan 2012–2021: A Strategic Plan for the U.S. Global Change Research Program,[8] USGCRP continues to build on its core interagency science capabilities with investments in science, tools, and data. Building on this strong scientific foundation, USGCRP is implementing the necessary long-term science investments to inform responses to high-priority societal impacts of global change, such as those related to changing patterns of weather extremes and potential thresholds and tipping points in human and natural systems.

USGCRP's vision is to help the Nation successfully meet the challenges of climate and global change. To help fulfill this vision, USGCRP will continue to conduct scientific research needed to advance our understanding of Earth system processes, to characterize past and current climate and global change, and to identify, understand, and better project associated impacts and risks. Moving forward, the Program will increasingly develop capacity and tools to help decision makers anticipate and better manage those risks. The FY 2014 priorities address challenges of immediate concern and are organized within three thematic areas:

- *Theme 1 – Extremes, Thresholds, and Tipping Points:* Improve observations, understanding and anticipation of the risks (and their confidence intervals) to human and natural systems from extremes, thresholds, and tipping points arising from climate-related environmental change.

- *Theme 2 – Integrated Research on Coupled Earth and Human Systems:* Enhance USGCRP's capacity to improve our understanding and integration of socioeconomic and biological aspects of global change into research and decision support.

- *Theme 3 – Actionable Science for Informed Policy Making and Management:* Strengthen the scientific basis for decision making and enhance accessibility and utility of data and tools for decision support at relevant scales (e.g., regional).

Applying scientific knowledge and information to practical, risk-based decisions requires the integration and translation of science into information that can be used by a diverse array of decision makers. In implementing its FY 2014 priorities, USGCRP will emphasize integration across the three themes highlighted above, in ways that connect science and research products to specific societal needs. These investments in global change research will support implementation of the actions laid out in the President's Climate Action Plan.

---

[8] **Website**: downloads.globalchange.gov/strategic-plan/2012/usgcrp-strategic-plan-2012.pdf

6 BUDGET SUPPLEMENT

The FY 2014 enacted budget for USGCRP programs is approximately $2.5 billion—an increase of about five percent over the 2013 operating level. This increase reflects the needs discussed formerly and represents a commitment by the Administration to ensure that USGCRP can fulfill its responsibilities under the law. Over the past five years, USGCRP's annual budget has averaged about $2.5 billion. It is important to remember that the budget crosscut table (Table 1) represents the funds self-identified by USGCRP agencies as their contributions to USGCRP. The budget crosscut does not include the costs of many agency investments that are directly relevant, and indeed necessary, to the ability of USGCRP to address national objectives related to climate and global change (e.g., many of the observing networks and satellite systems that are critical to documenting trends were originally implemented by their sponsoring agencies for current operational purposes, and those typically are not included in the budget crosscut).

Table 1: FY 2012–2014 U.S. Global Change Research Program Budget by Agency

Agency	FY 2012 Budget Enacted ($M)	FY 2013 Budget Operating ($M)	FY 2014 Budget Enacted ($M)
Department of Agriculture (USDA)	116	107	111
Department of Commerce (DOC)	319	301	329
Department of Energy (DOE)	211	209	217
Department of Health and Human Services (HHS)	14	10	8
Department of the Interior (DOI)	59	55	54
Department of Transportation (DOT)	1	1	1
Environmental Protection Agency (EPA)	18	17	18
National Aeronautics and Space Administration (NASA)	1427	1355	1431
National Science Foundation (NSF)	333	316	313
Smithsonian Institution (SI)	8	8	8
TOTALS	**2506**	**2379**	**2489**

Non-Add Agency	FY 2012 Budget Enacted ($M)	FY 2013 Budget Enacted ($M)	FY 2014 Budget Requested ($M)
Department of State (DOS)[1]	3	3	3
U.S. Agency for International Development (USAID)[1]	11	11	11

[1] USAID and DOS funding supports USGCRP and the Climate Change International Assistance effort. In the past, some of this funding was counted under both categories. These efforts do not add to the USGCRP total.

7 APPENDICES

7.1 USGCRP Member Agencies

This section summarizes the principal areas of focus related to global change research for each USGCRP member agency.

Department of Agriculture

USDA's global change research program empowers land managers, policy makers, and Federal agencies with science-based knowledge to manage the risks and opportunities posed by climate change; reduce GHG emissions; and enhance carbon sequestration. USDA's global change research program includes contributions from the Agricultural Research Service (ARS), the National Institute of Food and Agriculture (NIFA), the Forest Service (USDA-FS), Natural Resources Conservation Service (NRCS), National Agricultural Statistics Service (NASS), and Economic Research Service (ERS). This work is important to ensuring sustained food security for the Nation and the world; maintaining and enhancing forest and natural resource health; and identifying risks to agricultural production from temperature and precipitation changes, pests, invasive species, and disease.

Specifically, USDA conducts assessments and projections of climate-change impacts on agricultural and natural systems, and develops GHG inventories. USDA also develops cultivars, cropping systems, and management practices to improve drought tolerance and build resilience to climate variability. USDA promotes integration of USGCRP research findings into farm and natural resource management, and helps build resiliency to climate change by developing and deploying decision support. USDA maintains critical long-term data collection and observation networks, including the Snowpack Telemetry (SNOTEL) network, the Soil Climate Analysis Network (SCAN), the National Resources Inventory (NRI), and the Forest Inventory and Assessment (FIA).

Finally, USDA also engages in communication, outreach, and education through multiple forums, including its vast network of agricultural extension services.

Department of Commerce

NOAA and NIST comprise the Department of Commerce's (DOC's) participation in USGCRP.

NOAA's strategic climate goal is "an informed society anticipating and responding to climate and its impacts." NOAA's overall objective is to provide decision makers with a predictive understanding of the climate and to communicate climate information so that people can make more informed decisions in their lives, businesses, and communities. These outcomes are pursued by implementing a global observing system, conducting research to understand climate processes, developing improved modeling capabilities, and developing and deploying climate educational programs and information services. NOAA aims to achieve its climate goal through the following strategic objectives:

- Improved scientific understanding of the changing climate system and its impacts;

- Assessments of current and future states of the climate system that identify potential impacts and inform science, service, and stewardship decisions;

- Mitigation and adaptation efforts supported by sustained, reliable, and timely climate services; and

- A climate-literate public that understands its vulnerabilities to a changing climate and makes informed decisions.

NIST works with other Federal agencies to develop or extend internationally accepted traceable measurement standards, methodologies, and technologies that enhance measurement capabilities for science-based GHG emission inventories and measurements critical to advancing climate science research. NIST provides measurements and standards that support accurate, comparable, and reliable climate observations and provides calibrations and special tests to improve the accuracy of a wide range of instruments and techniques used in climate research and monitoring. In FY 2009, NIST was included as a discrete element of US-GCRP's budget crosscut to provide specific measurements and standards of direct relevance to the program.

Department of Defense

The Department of Defense (DOD)—while not supporting a formal mission dedicated to global change research—is developing policies and plans to manage and respond to the effects of climate change on DOD missions, assets, and the operational environment. Various research agencies within DOD sponsor and undertake basic research activities that concurrently satisfy both national security requirements as well as the strategic goals of USGCRP. These include the Office of Naval Research (ONR), the Air Force Office of Scientific Research (AFOSR), the Army Research Office (ARO), and the Defense Advanced Research Projects Agency (DARPA). When applicable, the research activities of these agencies are coordinated with other Federally sponsored research via USGCRP and other entities.

Because the performance of DOD systems and platforms are influenced by environmental conditions, understanding the variability of the Earth's environment and the potential for change is of great interest to the Department. The DOD is responsible for the environmental stewardship of hundreds of installations throughout the United States, and must continue incorporating geostrategic and operational energy considerations into force planning, requirements development, and acquisition processes. DOD relies on the Strategic Environmental Research and Development Program (SERDP), a joint effort among DOD, DOE, and EPA, to develop climate-change assessment tools and to identify the environmental variables that must be forecast with sufficient lead time to facilitate appropriate adaptive responses. Each service agency within

DOD incorporates the potential impact of global change into their long-range strategic plans. For example, the Navy's Task Force Climate Change (TFCC) assists in the development of science-based recommendations, plans, and actions to adapt to climate change. The USACE Engineer Research and Development Center (ERDC) Cold Regions Research and Engineering Laboratory (CRREL) also actively investigates the impacts of climate trends for DOD and other agencies. The CRREL research program responds to the needs of the military, but much of the research also benefits the civilian sector and is funded by non-military customers such as NSF, NOAA, NASA, DOE, and state governments.

Department of Energy

The Department of Energy's (DOE) Office of Science supports fundamental research to understand the energy-environment-climate connection and its implications for energy production, use, sustainability, and security—with particular emphasis on the potential impact of increased anthropogenic emissions. The ultimate goal is to advance a robust predictive understanding of Earth's climate and environmental systems and to inform the development of sustainable solutions to the Nation's energy and environmental challenges.

Two DOE research areas focus on areas of uncertainty in Earth systems models: Atmospheric System Research (science of aerosols, clouds, and radiative transfer); and Terrestrial Ecosystem Science (role of terrestrial ecosystems and carbon cycle observations). DOE also collaborates with NSF to develop the widely used Community Earth System Model, supports methods to obtain regional climate information, integrates analysis of climate-change impacts, and analyzes and distributes large climate datasets through the Program for Climate Model Diagnosis and Intercomparison and the Earth System Grid. The Department also supports the ARM Climate Research Facility, a scientific user facility that provides the research community with unmatched measurements permitting the most detailed high-resolution, three-dimensional documentation of evolving cloud, aerosol and precipitation characteristics in climate sensitive sites around the world.

Finally, DOE also conducts applied climate-related research through CCTP, which is centered in DOE's Office of Energy Policy and Systems Analysis and Office of Policy and International Affairs. CCTP develops and utilizes energy-economic models, including integrated assessment models, to evaluate policies and programs that enable cost-effective GHG reductions and accelerate the development and deployment of clean energy technologies. As part of this mission CCTP supports work to characterize climate-change impacts for use in policy analysis, vulnerability and adaptation assessment and agency rulemakings. DOE also conducts assessments of climate change on electric grid stability, water availability for energy production, and site selection of the next generation of renewable energy infrastructure.

Department of Health and Human Services

The U.S. Department of Health and Human Services (HHS) supports a broad portfolio of research and decision support initiatives related to environmental health and the health effects of global climate change, primarily through the National Institutes of Health (NIH) and the Centers for Disease Control (CDC). Research focuses on the need to better understand the vulnerabilities of individuals and communities to climate-related changes in health risks such as heat-related morbidity and mortality, respiratory effects of altered air contaminants, changes in transmission of infectious diseases, and impacts in the aftermath of severe weather events, among many others. Research efforts also seek to assess the effectiveness of various public health adaptation strategies to reduce climate vulnerability, as well as the potential health effects of interventions to reduce GHG emissions.

Specifically, HHS supports USGCRP by conducting fundamental and applied research on linkages between climate change and health, translating scientific advances into decision support tools for public health professionals, conducting ongoing monitoring and surveillance of climate-related health outcomes, and engaging the public health community in two-way communication about climate change.

Department of the Interior

USGS conducts global change research for the Department of the Interior (DOI) and comprises DOI's formal participation in USGCRP.

USGS scientists work with other agencies to provide policy makers and resource managers with scientifically valid information and predictive understanding of global change and its effects with the ultimate goal of helping the Nation understand, adapt to, and mitigate global change.

Specifically, the USGS Climate and Land Use Change Research and Development Program supports research to understand processes controlling Earth system responses to global change and model impacts of climate and land-cover change on natural resources. USGS geographic analyses and land remote-sensing programs (such as the Landsat satellite mission and the National Land Cover Database) provide data that is used to assess changes in land use, land cover, ecosystems, and water resources resulting from the interactions between human activities and natural systems. The science products and data sets from these programs are essential for conducting quantitative studies of carbon storage and GHG flux in the Nation's ecosystems.

USGS is also leading the establishment of regional DOI Climate Science Centers that will provide science and technical support to region-based partners dealing with the impacts of climate change on fish, wildlife, and ecological processes.

Department of State

Through the Department of State (DOS) annual funding, the United States is the world's leading financial contributor to the United Nations Framework Convention on Climate Change (UNFCCC) and to the IPCC—the principal international organization for the assessment of scientific, technical, and socioeconomic information relevant to the understanding of climate change, its potential impacts, and options for adaptation and mitigation. Recent DOS contributions to these organizations provide substantial support for global climate observation and assessment activities in developing countries. DOS also works with other agencies in promoting international co-operation in a range of bilateral and multilateral climate change initiatives and partnerships.

Department of Transportation

The Department of Transportation (DOT) conducts research to examine potential climate-change impacts on transportation, methods for increasing transportation efficiency, and methods for reducing emissions that contribute to climate change. DOT's Center for Climate Change and Environmental Forecasting coordinates transportation and climate-change research, policies, and actions within DOT and promotes comprehensive approaches to reduce emissions, address climate-change impacts, and develop adaptation strategies. DOT also contributes directly to USGCRP's National Climate Assessment through focused research such as the Center's Gulf Coast Studies.

The Federal Highway Administration, the Federal Transit Administration (FTA) and other DOT agencies are also undertaking climate impact and adaptation studies (including vulnerability and risk assessments), working with science agencies to develop regional climate data and projections, methodological research, supporting pilot programs, and providing assistance to transportation stakeholders including state and local agencies. The Federal Aviation Administration (FAA), for example, conducts research to support USGCRP by working with NASA, NOAA, and EPA in the Aviation Climate Change Research Initiative (ACCRI) to identify and address key scientific gaps regarding aviation climate impacts and inform mitigation.

Other DOT initiatives address climate change and improve the sustainability of the U.S. transportation sector including: The FAA and NASA manage the Continuous Lower Energy, Emissions, and Noise (CLEEN) program as a government industry consortium to develop technologies for energy efficiency, noise and emissions reduction, and alternative fuels; and FAA participates in the Commercial Aviation Alternative Fuels Initiative (CAAFI), a public-private coalition to encourage the development of alternative jet fuels.

Environmental Protection Agency

The core purpose of the Environmental Protection Agency's (EPA's) Global Change Research Program is to develop scientific information that supports stakeholders, policy makers, and society at large as they respond to climate change and associated impacts on human health, ecosystems, and socioeconomic systems. EPA's research is driven by the Agency's mission and statutory requirements, and includes: (1) improving scientific understanding of global change effects on air quality, water quality, ecosystems, and human health in the context of other stressors; (2) assessing and developing adaptation options to effectively respond to global change risks, increase resilience of human and natural systems, and promote their sustainability; and (3) developing an understanding of the potential environmental impacts and benefits of GHG emission reduction strategies to support sustainable mitigation solutions. This research is leveraged by EPA Program Offices and Regions to support mitigation and adaptation decisions and to promote communication with external stakeholders and the public.

EPA relies on USGCRP to develop high-quality scientific data and understanding about physical, chemical, and biological changes to the global environment and their relation to drivers of global change. EPA's Global Change Research Program connects these results to specific human and ecosystem health endpoints in ways that enable local, regional, and national decision makers to develop and implement strategies to protect human health and the environment. In turn, EPA's research provides USGCRP agencies with information about the connections between global change and local impacts and how local actions influence global changes.

Research activities include efforts to connect continental-scale temperature and precipitation changes to regional and local air quality and hydrology models to better understand the impacts of climate change on air quality and water quality, and to examine how watersheds will respond to large-scale climate and other global changes to inform decisions about management of aquatic ecosystems and expand understanding of the impacts of global change. Satellite and other observational efforts conducted by USGCRP are crucial to supporting EPA's efforts to understand how land use change, climate change, and other global changes are affecting watersheds and ecosystems, and the services they provide.

National Aeronautics and Space Administration

As stated in the 2010 National Space Policy, the National Aeronautics and Space Administration (NASA) plays a crucial role in conducting global change research, ensuring sustained monitoring capabilities, and advancing scientific knowledge of the global integrated Earth system through satellite observations and satellite system development. As such, NASA fully supports USGCRP's new *National Global Change Research Plan* to advance science, inform decisions, conduct sustain assessments, and communicate and educate. NASA actively contributes to USGCRP's National Climate Assessment and constitutes about half of the USGCRP budget.

NASA's global change activities have four integrated foci: satellite observations; technology development; research and analysis; and applications. Satellites provide critical global atmosphere, ocean, land, sea ice, and ecosystem measurements. NASA's seventeen on-orbit satellites measure numerous variables required to enhance understanding of Earth interactions. In 2013, NASA launched the Landsat Data Continuity Mission (LDCM) satellite to measure land cover and evapotranspiration and is developing other satellites for launch in 2014 and beyond. The Global Precipitation Measurement (GPM) mission launched in February 2014. GPM is an international satellite mission that will set a new standard for precipitation measurements from space, providing the next-generation observations of rain and snow worldwide every three hours. Applications projects extend the societal benefits of NASA's research, technology, and spaceflight programs to the broader U.S. public through the development and transition of user-defined tools for decision support.

NASA's program advances observing technology and leads to new and enhanced space-based observation and information systems. Science research and analysis of satellite observations and model results improve predictability and knowledge of the global integrated Earth system. Airborne

systems provide high resolution observations of variables relevant to global change research—including polar seas and ice sheets; atmospheric composition; carbon storage and flux in the Arctic; hurricanes in the Atlantic Ocean; and root-zone soil moisture at different locales in North America.

National Science Foundation

The National Science Foundation (NSF) addresses global change issues through investments that advance frontiers of knowledge, provide state-of-the-art instrumentation and facilities, develop new analytical methods, and enable cross-disciplinary collaborations while also cultivating a diverse, highly trained, workforce and developing educational resources. In particular, NSF global change programs support the research and related activities to advance fundamental understanding of physical, chemical, biological, and human systems and the interactions among them. The programs encourage interdisciplinary approaches to studying Earth system processes and the consequences of change, including how humans respond to changing environments and the impacts on ecosystems and the essential services they provide. NSF programs promote the development and enhancement of models to improve understanding of integrated Earth system processes and to advance predictive capability. NSF also supports fundamental research on the processes used by organizations and decision makers to identify and evaluate policies for mitigation, adaptation, and other responses to the challenge of a changing and variable environment. Long-term, continuous, and consistent observational records are essential for testing hypotheses quantitatively and are thus a cornerstone of global change research. NSF supports a variety of research observing networks that complement, and are dependent on, the climate monitoring systems maintained by its sister agencies.

NSF regularly collaborates with other USGCRP agencies to provide support for a range of multi-disciplinary research projects and is actively engaged in a number of international partnerships.

Smithsonian Institution

Within the Smithsonian Institution (SI), global change research is primarily conducted at the National Air and Space Museum, the National Museum of Natural History, the National Zoological Park, the Smithsonian Astrophysical Observatory, the Smithsonian Environmental Research Center, and the Smithsonian Tropical Research Institute. Research is organized around themes of atmospheric processes, ecosystem dynamics, observing natural and anthropogenic environmental change on multiple time scales, and defining longer term climate proxies present in the historical artifacts and records of the museums as well as in the geologic record. Most of these units participate in the Smithsonian Institution Global Earth Observatory (SIGEO) examining the dynamics of forests over decadal time frames.

The Smithsonian Grand Challenge Consortium for Understanding and Sustaining a Biodiverse Planet brings together researchers from around the Institution to focus on joint programs ranging from estimating volcanic emissions to ocean acidification measurement. Smithsonian paleontological research documents and interprets the history of terrestrial and marine ecosystems from 400 million years ago to the present. Other scientists study the impacts of historical environmental change on the ecology and evolution of organisms, including humans. Archaeobiologists examine the impact of early humans resulting from their domestication of plants and animals, creating the initial human impacts on planetary ecosystems.

These activities are joined by related efforts in the areas of history and art, such as the Center for Folklife and Cultural History, the National Museum of the American Indian, and the Cooper Hewitt Museum of Design to examine human responses to global change, within communities, reflected in art and culture, food, and music. Finally, Smithsonian outreach and education expands our scientific and social understanding of processes of change and represents them in exhibits and programs, including at the history and art museums of the Smithsonian. USGCRP funding enables the Smithsonian to leverage private funds for additional research and education programs on these topics.

U.S. Agency for International Development

The U.S. Agency for International Development (US-AID) supports programs that enable decision makers to apply high-quality climate information to decision making. USAID's climate change and development strategy calls for enabling countries to accelerate their transition to climate resilient, low emission sustainable economic development through direct programming and integrating climate change adaptation and mitigation objectives across the Agency's development portfolio. USAID is the lead contributor to bilateral assistance, with a focus on capacity building, civil society building, and governance programming, and creating the legal and regulatory environments needed to address climate change. USAID leverages scientific and technical resources from across the U.S. Government (e.g., NASA, NOAA, USDA, USGS) as it applies its significant technical expertise to provide leadership in development and implementation of low-emissions development strategies, creating policy frameworks for market-based approaches to emission reduction and energy sector reform, promoting sustainable management of agriculture lands and forests, and mainstreaming adaptation into development activities in countries most at risk. USAID has long-standing relationships with host country governments that enable it to work together to develop shared priorities and implementation plans. USAID's engagement and expertise in agriculture, biodiversity, infrastructure, and other critical climate sensitive sectors provide an opportunity to implement innovative cross-sectoral climate change programs. Finally, USAID bilateral programs work in key political and governance areas where multilateral agencies cannot.

7.2 Definitions

Adaptation: Adjustment in natural or *human systems* to a new or changing environment that exploits beneficial opportunities and moderates negative impacts.

Adaptation Science: Integrated scientific research that directly contributes to enabling adjustments in natural or human systems to a new or changing environment and that exploits beneficial opportunities or helps moderate negative effects.

Aerosols: Fine solid or liquid particles suspended in a gas. *Aerosols* may be of either natural or *anthropogenic* origin.

Anthropogenic: Resulting from or produced by human beings.

Assessments: Processes that involve analyzing and evaluating the state of scientific knowledge (and the associated degree of scientific certainty) and, in interaction with users, developing information applicable to a particular set of issues or decisions.

Atmosphere: The gaseous envelope surrounding Earth.

Biodiversity: The total diversity of all organisms and ecosystems at various spatial scales.

Biosphere: The part of the Earth system comprising all ecosystems and living organisms, in the *atmosphere*, on land, or in the ocean, including derived dead organic matter, such as litter, soil organic matter, and oceanic detritus*.*

Carbon cycle: The term used to describe the flow of carbon (in various forms, e.g., as carbon dioxide, calcium carbonate) through the *atmosphere, ocean*, terrestrial *biosphere*, and *lithosphere.*

Carbon sequestration: The process of increasing the carbon content of a carbon reservoir other than the atmosphere.

Climate: The mean and variability of relevant measures of the *atmosphere*-ocean system over periods ranging from decades to thousands or millions of years.

Climate change: A statistically significant variation in either the mean state of the *climate* or in its *variability*, persisting for an extended period (typically decades or longer). Climate change may be due to natural internal processes or to external forcing, including changes in solar radiation and volcanic eruptions, or to persistent human-induced changes in atmospheric composition or in land use. *See also climate variability.*

Climate model: A numerical representation of the *climate system* based on the mathematical equations governing the physical, chemical, and biological properties of its components and including treatment of key physical processes and interactions, cast in a form suitable for numerical approximation making use of computers.

Climate projection: A projection of the response of the climate system to emission or concentration scenarios of *greenhouse gases* or *aerosols*, or *radiative forcing scenarios*, often based upon simulations by *climate models*. Climate projections are distinguished from climate predictions in order to emphasize that *climate projections* depend upon the *emission*/concentration/ radiative forcing scenario used, which are based on assumptions concerning, for example, future socio-economic and technological developments that may or may not be realized and are therefore subject to substantial uncertainty.

Climate system: The highly complex system consisting of five major components: the *atmosphere*, the hydrosphere, the cryosphere, the land surface, and the *biosphere*, and the interactions among them.

Climate variability: Variations in the mean state and other statistics (such as the occurrence of extremes, etc.) of the *climate* on all temporal and spatial scales beyond that of individual weather events. These variations are often due to internal processes within the *climate system* (internal variability), or to variations in natural or *anthropogenic* external forcing (external variability).

Committee on Environment, Natural Resources, and Sustainability (CENRS): A subcommittee of the *National Science and Technology Council* (NSTC) established to assist the NSTC in increasing the overall productivity and application of Federal research and development efforts in the areas of environment, natural resources, and sustainability, and to provide a formal mechanism for interagency coordination in these areas. CENRS encompasses the *Subcommittee on Global Change Research*, the steering committee of the *U.S. Global Change Research Program.*

Decision support: The provision of timely and useful information that addresses specific questions.

Earth system: The unified set of physical, chemical, biological, and social components, processes, and interactions that together determine the state and dynamics of planet Earth.

Ecosystem: A system of living organisms interacting with each other and their physical environment as an ecological unit.

Ecosystem services: The conditions and processes through which natural ecosystems, and the species that make them up, sustain and fulfill human life. Examples include provision of clean water, maintenance of liveable climates, pollination of crops and native vegetation, and fulfillment of people's cultural, spiritual, intellectual needs.

Emissions: In the *climate change* context, emissions refer to the release of radiatively or chemically active substances (e.g., *greenhouse gases* and/or their precursors, *aerosols*) into the *atmosphere* over a specified area and period of time.

Extreme weather event: An event that is rare at a particular place and time of year. Definitions of "rare" vary, but an extreme *weather* event would normally be as rare as or rarer than the 10th or 90th percentile of the observed probability density of *weather* events.

Feedback: An interaction mechanism between processes such that the result of an initial process triggers changes in a second process and that in turn influences the initial one. A positive feedback intensifies the original process, and a negative feedback reduces it.

Fiscal Year (FY): A period used for calculating annual ("yearly") financial statements in the Federal Government.

General Circulation (GCM) or Atmosphere/Ocean Global Climate Model: A numerical representation of the *climate system* based on the physical and chemical properties of its components, their interactions and feedback processes, and accounting for all or some of its known properties.

Global change: Changes in the global environment (including alterations in *climate*, land productivity, oceans or other water resources, atmospheric composition and/or chemistry, and ecological systems) that may alter the capacity of the Earth to sustain life.

Global Change Information System: An information system that establishes repositories of *climate* and *global change* information and interoperable interfaces to agency data centers so that data can be easily and efficiently accessed, integrated, maintained over time and expanded as needed into the future.

Global change research: Study, monitoring, assessment, prediction, and information management activities to describe and understand the interactive physical, chemical, and biological processes that regulate the total *Earth system*; the unique environment that the Earth provides for life; changes that are occurring in the *Earth system*; and the manner in which such system, environment, and changes are influenced by human actions.

Global Change Research Act (GCRA; Section 102, P. L. 101–606): A 1990 act establishing the *U.S. Global Change Research Program*, an interagency program aimed at understanding and responding to global change, including the cumulative effects of human activities and natural processes on the environment, to promote discussions toward international protocols in global change research, and for other purposes.

Greenhouse gas (GHG): Any gas that absorbs infrared radiation (heat) in the *atmosphere*. Greenhouse gases include, but are not limited to, water vapor, carbon dioxide, methane, nitrous oxide, chlorofluorocarbons, hydrochlorofluorocarbons, ozone, hydrofluorocarbons, perfluorocarbons, and sulfur hexafluoride.

Human system: Any system in which human organizations play a major role. Often, but not always, the term is synonymous with 'society' or 'social system' e.g., agricultural system, political system, technological system, economic system.

Human-natural system: Integrated systems in which human and natural components interact, such as the interaction between socioeconomic and biophysical processes in urban ecosystems.

In situ: Measurements obtained through instruments that are in direct contact with the subject (e.g., a soil thermometer), as opposed to those collected by remote instruments (e.g., a radar altimeter).

Integrated Assessment Models: A method of analysis that combines results and models from the physical, biological, economic, and social sciences, and the interactions between these components, in a consistent framework, to evaluate the status and consequences of environmental change and the policy responses to it.

Intergovernmental Panel on Climate Change (IPCC): An international scientific body for the assessment of *climate change*, established by the United Nations Environmental Programme and the United Nations World Meteorological Organization.

IPCC AR5: The fifth in a series of assessment reports by the *Intergovernmental Panel on Climate Change*, intended to assess the socioeconomic aspects of climate change and implications for sustainable development, risk management, and the framing of a response through both *adaptation* and *mitigation*.

Land cover: The land surface covering, including areas of vegetation (forests, shrub lands, crops, deserts, lawns), bare soil, developed surfaces (paved land, buildings), and wet areas and bodies of water (watercourses, wetlands).

Landsat Program: The Landsat Program is a series of Earth-observing satellite missions jointly managed by NASA and USGS.

Land use: The total of arrangements, activities, and inputs undertaken in a certain land cover type (a set of human actions). The term land use is also used in the sense of the social and economic purposes for which land is managed (e.g., grazing, timber extraction, conservation).

Land use and land cover change: A change in the use or management of land by humans that may lead to a change in land cover.

Metadata: Information about data concerning how and when they were measured, their quality, known problems, and other characteristics.

Mitigation (climate change): An intervention to reduce the sources or enhance the sinks of greenhouse gases and other climate forcing agents. This intervention could include approaches devised to reduce *emissions* of *greenhouse gases* to the atmosphere, or to enhance their removal from the *atmosphere* through storage in geological formations, soils, biomass, or the ocean.

Monitoring: A scientifically designed system of continuing standardized measurements and observations and the evaluation thereof. Monitoring is specifically intended to continue over long time periods.

National Academy of Sciences (NAS): An honorific society of distinguished scholars engaged in scientific and engineering research established by an Act of Congress in 1863, which calls upon the NAS to "investigate, examine, experiment, and report upon any subject of science or art" whenever called upon to do so by any department of the Government.

National Climate Assessment (NCA): An assessment conducted under the auspices of the Global Change Research Act of 1990, which requires a report to the President and the Congress every four years that evaluates, integrates, and interprets the findings of the *U.S. Global Change Research Program*.

National Science and Technology Council (NSTC): A Cabinet-level Council established by Executive Order 12881 that is the principal means within the executive branch to coordinate science and technology policy across the diverse entities that make up the Federal research and development enterprise.

Observations: Measurements (either continuing or episodic) of variables in *climate* and related systems.

Observing system: A coordinated series of instruments for long-term observations of the land surface, *biosphere*, solid Earth, *atmosphere*, and/or oceans to improve understanding of Earth as an integrated system.

Ocean acidification: The phenomenon in which the pH of the ocean becomes more acidic due to increased levels of carbon dioxide in the *atmosphere*, which, in turn, increase the amount of dissolved carbon dioxide in seawater. Ocean acidification may lead to reduced calcification rates of calcifying organisms such as corals, mollusks, algae and crustacea.

Office of Science and Technology Policy (OSTP): A division of the Executive Office of the President (EOP) established by Congress in 1976 with a broad mandate to advise the President and others within the EOP on the effects of science and technology on domestic and international affairs. The 1976 Act also authorizes OSTP to lead interagency efforts to develop and implement sound science and technology policies and budgets, and to work with the private sector, state and local governments, the science and higher education communities, and other nations toward this end.

Ozone: A very active colorless gas consisting of three atoms of oxygen, readily reacting with many other substances.

Prediction: A probabilistic description or forecast of a future outcome based on *observations* of past and current conditions and quantitative models (e.g., a prediction of an El Niño event).

Projection: A description of the response of a natural system (e.g., the *climate system*) to forcing (e.g., *radiative forcing*) at an assumed future level. Climate *projections* are distinguished from climate *predictions* in order to emphasize that climate *projections* depend on *scenarios* of future socioeconomic, technological, and policy developments that may or may not be realized.

Radiative forcing: A process that directly changes the average energy balance of the Earth-atmosphere system by affecting the balance between incoming solar radiation and outgoing radiation. A positive forcing warms the surface of the Earth and a negative forcing cools the surface.

Scenario: A coherent description of a potential future situation that serves as input to more detailed analyses or modeling. Scenarios are tools that explore, "if…, then…." statements, and are not *predictions* of or prescriptions for the future.

Stakeholders: Individuals or groups whose interests (financial, cultural, value-based, or other) are affected by *climate variability, climate change,* or options for *adapting* to or *mitigating* these phenomena. Stakeholders are important partners with the research community for development of decision support resources.

Storm surge: The temporary increase, at a particular locality, in the height of the sea due to extreme meteorological conditions (low atmospheric pressure and/or strong winds).

Subcommittee on Global Change Research (SGCR): The steering committee of the *U.S. Global Change Research Program* (USGCRP) under the *Committee on Environment, Natural Resources, and Sustainability,* overseen by the Executive Office of the President. SGCR is composed of representatives from each of the member agencies of USGCRP.

Sustainability: Balancing the needs of present and future generations while substantially reducing poverty and conserving the planet's life support systems.

System: Integration of interrelated, interacting, or interdependent components into a complex whole.

Technology: An approach, including both the experimental technique and the instrumental and scientific infrastructure needed to implement it.

Tipping point: A critical *threshold* at which a tiny perturbation can qualitatively alter the state or development of a system.

Threshold: A point in a system after which any change that is described as abrupt is one where the change in the response is much larger than the change in the forcing. The changes at the threshold are therefore abrupt relative to the changes that occur before or after the threshold and can lead to a transition to a new state.

Uncertainty: An expression of the degree to which a value (e.g., the future state of the *climate system*) is unknown. Uncertainty in future *climate* arises from imperfect scientific understanding of the behavior of physical systems, and from inability to predict human behavior.

United Nations Framework Convention on Climate Change (UNFCCC): The United Nations Framework Convention on Climate Change is an international environmental treaty produced at the United Nations Conference on Environment and Development (UNCED) intended to stabilize *greenhouse gas* concentrations in the atmosphere at a level that would prevent dangerous anthropogenic interference with the *climate system.*

U.S. Global Change Research Program (USGCRP): An interagency program that coordinates and integrates Federal research on changes in the global environment and their implications for society. USGCRP began as a presidential initiative in 1989 and was mandated by Congress in the Global Change Research Act of 1990 (P.L. 101–606). Thirteen departments and agencies participate in USGCRP. The program is steered by the *Subcommittee on Global Change Research* under the *Committee on Environment and Natural Resources*, overseen by the Executive Office of the President, and facilitated by a National Coordination Office (NCO).

Vulnerability: The degree to which a system is susceptible to, or unable to cope with, the effects of changing conditions, such as *climate* and *global change* (including *climate variability* and *extremes*, as well as *climate change* in conjunction with other stressors).

Weather: The specific condition of the *atmosphere* at a particular place and time. It is measured in terms of parameters such as wind, temperature, humidity, atmospheric pressure, cloudiness, and precipitation.

7.3 Glossary of Acronyms

ACCRI – Aviation Climate Change Research Initiative

AFOSR – Air Force Office of Scientific Research

AR5 – IPCC Fifth Assessment Report

ARO – Army Research Office

ARS – Agricultural Research Service

CAAFI – Commercial Aviation Alternative Fuels Initiative

CDC – Centers for Disease Control and Prevention

CENRS – Committee on Environment, Natural Resources, and Sustainability

CEQ – Council on Environmental Quality

CLEEN – Continuous Lower Energy, Emissions, and Noise

CMIP5 – Fifth-phase Coupled Model Intercomparison Project

CRREL – Cold Regions Research and Engineering Laboratory

DOD – Department of Defense

DOE – Department of Energy

DOI – Department of the Interior

DOS – Department of State

DOT – Department of Transportation

EPA – Environmental Protection Agency

ERDC – Engineer Research and Development Center

FAA – Federal Aviation Administration

FIA – Forest Inventory Analysis

FY – Fiscal Year

GCIS – Global Change Information System

GCRA – Global Change Research Act

GHG – Greenhouse Gas

HHS – U.S. Department of Health and Human Services

IAM – Integrated Assessment Models

IPCC – Intergovernmental Panel on Climate Change

LDCM – Landsat Data Continuity Mission

MATCH – Metadata Access Tool for Climate and Health

NAS – National Academy of Sciences

NASA – National Aeronautics and Space Administration

NASS – National Agricultural Statistics Service

NCA – National Climate Assessment

NCO – National Coordination Office

NEO – National Earth Observations

NIFA – National Institute of Food and Agriculture

NIH – National Institutes of Heath

NMME – North American Multi-Model Ensemble

NOAA – National Oceanic and Atmospheric Administration

NRCS – Natural Resources Conservation Service

NRI – National Resources Inventory

NSF – National Science Foundation

NSTC – National Science and Technology Council

ONR – Office of Naval Research

OSTP – Office of Science and Technology Policy

RISA – Regional Integrated Sciences and Assessments

SCAN – Soil Climate Analysis Network

SERDP – Strategic Environmental Research and Development Program

SGCR – Subcommittee for Global Change Research

SIGEO – Smithsonian Institution Group on Earth Observations

SNOTEL – Snowpack Telemetry

STRI – Smithsonian Tropical Research Institute

TFCC – Task Force Climate Change

UN – United Nations

UNEP – United Nations Environment Programme

UNFCCC – United Nations Framework Convention on Climate Change

USACE – U.S. Army Corps of Engineers

USAID – U.S. Agency for International Development

USDA – U.S. Department of Agriculture

USGCRP – U.S. Global Change Research Program

USGS – U.S. Geological Survey

WMO – World Meteorological Organization

www.ingramcontent.com/pod-product-compliance
Lightning Source LLC
Chambersburg PA
CBHW080616180526
45168CB00007B/2930